T0176738

An Introduction to Surface Analysis by XPS and AES

# An Introduction to Surface Analysis by XPS and AES

*John F. Watts*
University of Surrey, UK

*John Wolstenholme*
Crowborough, UK

Second Edition

This edition first published 2020


"First edition – John Wiley & Sons Ltd, 2003".



*Registered Offices*
John Wiley & Sons, Inc., 111 River Street, Hoboken, NJ 07030, USA
John Wiley & Sons Ltd, The Atrium, Southern Gate, Chichester, West Sussex, PO19 8SQ, UK

*Editorial Office*
The Atrium, Southern Gate, Chichester, West Sussex, PO19 8SQ, UK

For details of our global editorial offices, customer services, and more information about Wiley products visit us at www.wiley.com.

Wiley also publishes its books in a variety of electronic formats and by print-on-demand. Some content that appears in standard print versions of this book may not be available in other formats.



*Library of Congress Cataloging-in-Publication Data*

Names: Watts, John F., author. | Wolstenholme, John, author.
Title: An introduction to surface analysis by XPS and AES / John F. Watts, The Surface Analysis Laboratory, Department of Mechanical Engineering Sciences, University of Surrey, John Wolstenholme.
Description: Second edition. | Hoboken : Wiley, 2020. | Includes bibliographical references and index.
Identifiers: LCCN 2019023206 (print) | LCCN 2019023207 (ebook) | ISBN 9781119417583 (cloth) | ISBN 9781119417620 (adobe pdf) | ISBN 9781119417644 (epub)
Subjects: LCSH: Surfaces (Technology)–Analysis. | Electron spectroscopy. | Photoelectron spectroscopy. | Auger effect.
Classification: LCC TP156.S95 W373 2020 (print) | LCC TP156.S95 (ebook) | DDC 620/.44–dc23
LC record available at https://lccn.loc.gov/2019023206
LC ebook record available at https://lccn.loc.gov/2019023207

Cover Design: Wiley
Cover Image: Courtesy of Robin Simpson, Thermo Fisher Scientific, Background © Ralf Hiemisch/Getty Images

Set in 10/12pt Warnock by SPi Global, Pondicherry, India
Printed and bound in Singapore by Markono Print Media Pte Ltd

10 9 8 7 6 5 4 3 2 1

# Contents

## Preface to First Edition

When one of us (JFW) wrote an earlier introductory text in electron spectroscopy the aim was to fill a gap in the market of the time (1990) and produce an accessible text for undergraduates, first-year postgraduates, and occasional industrial users of XPS and AES. In the intervening years the techniques have advanced in both the area of use and, particularly, in instrument design. In XPS, X-ray monochromators are now becoming the norm and imaging has become commonplace. In AES, field emission sources are to be seen on high-performance systems. Against that backdrop it was clear that a new, broader introductory book was required that explored the basic principles and applications of the techniques, along with the emerging innovation in instrument design.

We hope that this book has achieved that aim and will be of use to newcomers to the field, both as a supplement to undergraduate and master's level lectures, and as a stand-alone volume for private study. The reader should obtain a good working knowledge of the two techniques (although not, of course, of the operation of the spectrometers themselves) in order to be able to hold a meaningful dialogue with the provider of an XPS or AES service at, for example, a corporate research laboratory or service organisation.

Further information on all the topics can be found in the Bibliography and the titles of papers and so on have been included along with the more usual citations to guide such reading. The internet provides a valuable resource for those seeking guidance on XPS and AES and rather than attempt to be inclusive in our listing of such sites we merely refer readers to the UKSAF site and its myriad of links. Finally, we have both been somewhat perturbed by the degree of confusion and sometimes contradictory definitions regarding some of the terms used in electron spectroscopy. In an attempt to clarify the situation, we have included a glossary of the more common terms. This has been taken from ISO 18115 and we thank ISO for permission to reproduce this from their original document.

*John F. Watts*
Guildford, Surrey, UK
*John Wolstenholme*
East Grinstead, West Sussex, UK

# Preface to Second Edition

It is now some 16 years since the publication of the first edition of this book. We have both been gratified by the positive manner in which it has been received by the community, testament to the need for an introductory text written by experienced practitioners. It gives us much satisfaction to see it in research and contract labs across the globe and frequently referenced in undergraduate and postgraduate reports. It thus appears to have fulfilled the perceived need outlined in the Preface to the first edition.

This perhaps begs the question 'why wait sixteen years to produce the second edition?' The answer to this is probably familiar to many practicing scientists and engineers in that the plan had informally been in place for several years, as had a formal contract; it was merely a question of allocating time in busy schedules.

The aim of the text remains the same, as enunciated in the first edition, and we have retained the same overall format. Basic principles remain, but instrumentation is constantly changing and improving as manufacturers seek better and better performance and functionality from their products. We have tried to reflect such developments in the text and make sure that it is as up to date as possible. In particular, we deal with a number of recent innovations in a manner accessible to newcomers to surface analysis. These include:

- Gas cluster ion sources.
- Newly developed high-energy X-ray sources (CrKα and GaKα) that are now available on commercial XPS systems.
- Near ambient pressure XPS, (a technique which is in its infancy but where the UK is arguably developing a lead).
- Fully automated, entry-level, high-performance systems.

Certain areas, such as sample preparation, sample mounting, quantification, peak fitting, energy loss background analysis, multi-technique analysis and multi-technique profiling are treated in more detail than before.

The year of publication (2019) is the Golden Jubilee of the launch of XPS and AES as commercially available analysis methods. It is a rather salutary thought that both of us have been involved with applied surface analysis for more than three quarters of this time, which gives us cause to reflect on the many innovations that have taken place during this time. As a celebration of 50 years of XPS we include images of one of the first commercial XPS systems and a sectioned analyser from such a system, overleaf.

As they say at all Golden Jubilee celebrations 'Here's to the next fifty years!'

*John F. Watts*
Guildford, Surrey, UK
*John Wolstenholme*
Crowborough, East Sussex, UK

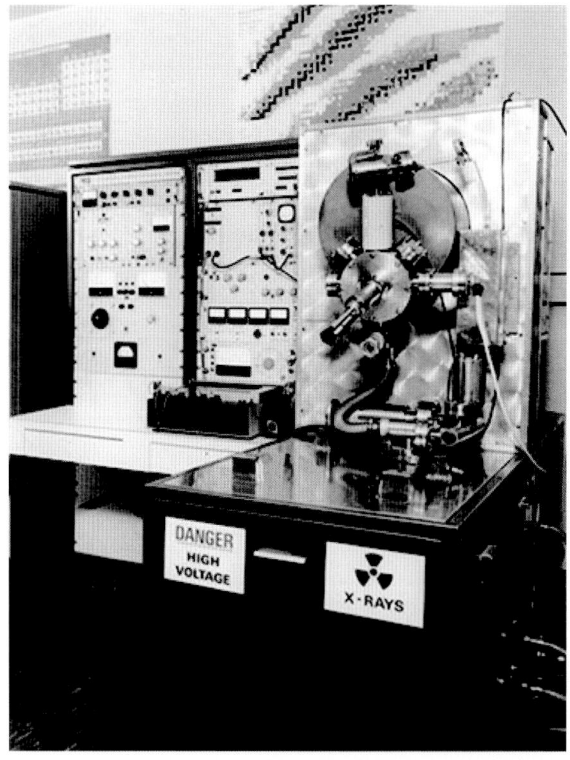


The first commercial XPS system from VG Scientific (now Thermo Fisher Scientific), a 1969 ESCA 2 electron spectrometer, installed at the University of Surrey

Sectioned concentric hemispherical analyser from the above spectrometer.

# Acknowledgements

There are many who have contributed ideas and data to this book; current and former colleagues and students in particular. Specifically, those in The Surface Analysis Laboratory at the University of Surrey (Drs Marie-Laure Abel, Simon Bacon, Jorge Bañuls Ciscar, Rossana Grilli, Chris Mallinson, Sabrina Tardio) and Drs Richard White, Tim Nunney and the staff at Thermo Fisher Scientific, have provided much new data to replace or complement that provided for the first edition. Dr Andreas Thissen at SPECS Surface Nano Analysis GmbH provided information and examples related to near ambient pressure XPS. In this, the golden jubilee year of XPS and AES, most of the innovators associated with XPS in the early days are no longer active. One notable exception being Professor Jim Castle who continues to edit a journal, publish papers, sit on ISO/TC201 and present work at conferences, in addition to being a source of inspiration to all associated with The Surface Analysis Laboratory at the University of Surrey.

Certain figures and data have been reproduced from other sources and we thank the copyright holder's for their permission to do so. In particular, the Wiley journal ***Surface and Interface Analysis*** has provided many valuable illustrations of the application of the techniques of XPS and AES. The cover design makes use of original graphics by Dr Robin Simpson (Thermo Fisher Scientific) and features the Nexsa™ automated XPS system.

The following figures are based on the data and other material provided by these organizations:

| Organization | Figures |
|---|---|
| JEOL Ltd, 3-1-2 Musashino, Akishima, Tokyo 196-8558, Japan. | 2.1, 2.9, 7.24 |
| Physical Electronics USA Inc. 18725 Lake Drive East, Chanhassen, Minnesota, USA | 7.17 (Sample courtesy of Fujifilm Corp), 7.25, 7.26 |

*(Continued)*

| Organization | Figures |
| --- | --- |
| Scienta Omicron AB, P.O. Box 15120, 750 15 Uppsala, Sweden | 2.3 |
| SPECS Surface Nano Analysis GmbH, Voltastrasse 5, Berlin, Germany | 2.19, 2.20, 7.59, 7.60 |
| Thermo Fisher Scientific Inc., Unit 19, The Birches Industrial Estate, East Grinstead RH19 1XZ UK | 1.1, 1.4, 1.6, 1.7, 1.8, 2.2, 2.6, 2.8, 2.11, 2.15, 2.17, 2.25, 3.2, 3.4, 3.6, 3.8, 3.12, 3.13, 4.2, 4.4, 4.5, 4.6, 4.7, 4.8, 4.9, 4.10, 4.11, 4.12, 4.14, 4.15, 4.22, 4.23, 4.24, 4.26, 4.27, 4.28, 5.1, 5.4, 5.6, 5.7, 5.8, 5.9, 5.10, 6.1, 6.2, 7.1, 7.19, 7.27, 7.28, 7.29, 7.30, 7.31, 7.32, 7.33, 7.34, 7.35, 7.37, 7.38, 7.39, 7.50, 7.62, 7.63 |

# 1

# Electron Spectroscopy: Some Basic Concepts

## 1.1 Analysis of Surfaces

All solid materials interact with their surroundings through their surfaces. The physical and chemical composition of these surfaces determine the nature of the interactions. Their surface chemistry will influence such factors as corrosion rates, catalytic activity, adhesive properties, wettability, contact potential, failure mechanisms, etc. Surfaces, therefore, influence many crucially important properties of the solid.

Despite the undoubted importance of surfaces, only a very small proportion of the atoms of most solids are found at the surface. Consider, for example, a 1 cm cube of a typical transition metal (e.g. nickel). The cube contains about $9 \times 10^{22}$ atoms of which about $6 \times 10^{15}$ are at the surface. The proportion of surface atoms is therefore approximately 1 in $10^8$ or 10 ppb. If we want to detect impurities at the nickel surface at a concentration of 0.1% then we need to detect materials at a concentration level of 0.01 ppb within the cube. The exact proportion of atoms at the surface will depend upon the shape and surface roughness of the material as well as its composition. The above figures simply illustrate that a successful technique for analysing surfaces must have at least two characteristics:

a) It must be extremely sensitive.
b) It must be efficient at filtering out signal from the vast majority of the atoms present in the sample.

This book is largely concerned with X-ray photoelectron spectroscopy (XPS) and Auger electron spectroscopy (AES). As will be shown, each of these techniques has the required characteristics but, in addition, they can answer other important questions:

a) Which elements are present in the near-surface region of a solid?
b) Which chemical states of these elements are present?

*An Introduction to Surface Analysis by XPS and AES*, Second Edition.
John F. Watts and John Wolstenholme.

c) How much of each chemical state of each element is present?
d) What is the spatial distribution of the materials in the near surface region in three-dimensions and how does that vary with time?
e) If material is present as a thin film at the surface:
   i) How thick is the film?
   ii) How uniform is the thickness?
   iii) How uniform is the chemical composition of the film?

In electron spectroscopy, we are concerned with the emission and energy analysis of low energy electrons, usually in the range 20–2000 eV[1] (the use of X-ray anodes that generate X-rays having a photon energy much higher than 2000 eV is becoming more popular). These electrons are liberated from the sample being examined as a result of the photoemission process (in XPS) or the radiationless de-excitation of an ionised atom by the Auger emission process in AES and scanning Auger microscopy (SAM). The distinction between AES and SAM is worthy of consideration. AES is a broad term that implies excitation of Auger electrons using a beam of electrons but makes no claim to be a technique that features high spatial resolution. SAM, on the other hand, always makes use of a finely focussed electron beam, typically in the range 10–100 nm, and provides results in the form of spatially resolved images derived from Auger electron data.

In the simplest terms, an electron spectrometer consists of the sample under investigation, a source of primary radiation, and an electron energy analyser all contained within a vacuum chamber, preferably operating in the ultra-high vacuum (UHV) regime. In practice, there will often be a secondary UHV chamber fitted with various sample preparation facilities and perhaps ancillary analytical facilities. A data system will be used for data acquisition and subsequent processing. The source of the primary radiation for the two methods is different; XPS making use of soft X-rays, most commonly monochromated Al Kα X-rays, although a twin anode arrangement is still often used (the most popular being Al Kα combined with Mg Kα), whereas AES and SAM rely on the use of an electron gun. The specification for electron guns used in Auger analysis varies tremendously, particularly as far as the spatial resolution is concerned, which, for finely focused guns, may be between 5 μm and < 10 nm. In principle, the same energy analyser may be used for both XPS and AES; consequently, the two techniques are often to be found in the same analytical instrument.

Before considering the uses and applications of the two methods, it is helpful to review the physical principles of the two processes along with their strengths and weaknesses.

---

1 Units: In electron spectroscopy, energies are usually expressed in the non-SI unit 'electron volt (eV)' which is a unit of energy equal to the work done on an electron in accelerating it through a potential difference of 1 V. 1 eV ≈ $1.6 \times 10^{-19}$ J.

## 1.2 Notation

XPS and AES measure the energy of electrons emitted from a material. It is necessary therefore to have some formal way to describe which electrons are involved with each of the observed transitions. The notation used in XPS is different from that used in AES. XPS uses the so-called spectroscopists' or chemists' notation while Auger electrons are identified by the X-ray notation.

### 1.2.1 Spectroscopists' Notation

In this notation, the photoelectrons observed are described by means of their quantum numbers. Transitions are usually labelled according to the scheme $nl_j$.

The first part of this notation is the principal quantum number, *n*. This takes integer values of 1, 2, 3, etc. The second part of the nomenclature, *l*, is the quantum number which describes the orbital angular momentum of the electron. This takes integer values 0, 1, 2, 3, etc. However, this quantum number is usually given a letter rather than a number as shown in Table 1.1.

The peaks in XPS spectra derived from orbitals whose angular momentum quantum number is greater than 0 are usually split into two. This is a result of the interaction of the electron angular momentum due to its spin with its orbital angular momentum. Each electron has a quantum number associated with its spin angular momentum[2], s. The value of s can be either +½ or −½. The two angular momenta are added vectorially to produce the quantity j in the expression $nl_j$, i.e. $j = |l + s|$. Thus, an electron from a p orbital can have a j value of ½ (l-s) or $^3/_2$ (l + s); similarly, electrons from a d orbital can have j values of either $^3/_2$ or $^5/_2$. The relative intensity of the components of the doublets formed by the spin orbit coupling is dependent upon their relative populations (degeneracies) which are given by the expression (2j + 1) so, for an electron from a d

Table 1.1 Notation given to the quantum numbers which describe orbital angular momentum.

| Value of *l* | Usual notation |
|---|---|
| 0 | s |
| 1 | p |
| 2 | d |
| 3 | f |

2 The electron spin quantum number, s, should not be confused with the description of the orbitals whose angular momentum is equal to zero were the S is normally given in upper case.

Table 1.2 The relationship between quantum numbers, spectroscopists' notation and X-ray notation.

| Quantum numbers | | | | |
|---|---|---|---|---|
| n | l | s | Spectroscopy notation | X-ray notation |
| 1 | 0 | ±½ | $1s_{1/2}$[a] | K |
| 2 | 0 | ±½ | $2s_{1/2}$[a] | $L_1$ |
| 2 | 1 | +½ | $2p_{1/2}$ | $L_2$ |
| 2 | 1 | ±½ | $2p_{3/2}$ | $L_3$ |
| 3 | 0 | ±½ | $3s_{1/2}$[a] | $M_1$ |
| 3 | 1 | ±½ | $3p_{1/2}$ | $M_2$ |
| 3 | 1 | ±½ | $3p_{3/2}$ | $M_3$ |
| 3 | 2 | ±½ | $3d_{3/2}$ | $M_4$ |
| 3 | 2 | ±½ | $3d_{5/2}$ | $M_5$ |
| | | | | etc. |

[a] usually identified as 1s, 2s, 3s, in XPS, the subscript is omitted.

orbital, the relative intensities of the 3/2 and 5/2 peaks are 2 : 3. The spacing between the components of the doublets depends upon the strength of the spin orbit coupling. For a given value of both n and l, the separation increases with the atomic number of the atom. For a given atom, it decreases both with increasing n and with increasing l.

### 1.2.2 X-ray Notation

In X-ray notation, the principal quantum numbers are given letters *K, L, M,* etc. while subscript numbers refer to the j values described above. The relationship between the notations is given in Table 1.2.

As will be seen later, the Auger process involves three electrons and so the notation must take account of this. This is done simply by listing the three electrons, for example a peak in an Auger spectrum may be labelled $KL_1L_3$ or $L_2M_5M_5$. For convenience, the subscripts are sometimes omitted.

## 1.3 X-ray Photoelectron Spectroscopy

In XPS we are concerned with a special form of photoemission, i.e. the ejection of an electron from a core level caused by an X-ray photon of energy $h\nu$. The energy of the emitted photoelectron is then analysed by the electron spectrometer and the data presented as a graph of intensity (usually expressed as counts or counts per second) versus electron energy; the X-ray induced photoelectron spectrum.

The kinetic energy ($E_K$) of the electron is the experimental quantity measured by the spectrometer, but this is dependent on the energy of the X-ray source employed and is therefore not an intrinsic material property. The binding energy of the electron ($E_B$) is the parameter which identifies the electron specifically, both in terms of its parent element and atomic energy level. The relationship between the parameters involved in the XPS experiment is as follows:

$$E_B = h\nu - E_K - \phi$$

Where $h\nu$ is the photon energy, $E_K$ is the kinetic energy of the electron, and $\phi$ is the work function of the spectrometer.

As all three quantities on the right-hand side of the equation are known or measurable, it is a simple matter to calculate the binding energy of the electron. In practice, this task will be performed by the control electronics or data system associated with the spectrometer and the operator merely selects a binding or kinetic energy scale whichever is considered the more appropriate. It should be emphasised that the positions of XPS peaks on the binding energy scale are independent of the X-ray photon energy but the positions of X-ray induced Auger peaks will depend upon the photon energy. On the other hand, if the spectrum is displayed on a kinetic energy scale then the position of the Auger peaks will remain constant while the position of the XPS peaks will change with photon energy.

Figure 1.1a is a schematic representation of the arrangement of the electrons in a carbon atom as a 1s electron is being ejected following an interaction with an X-ray photon. The carbon 1s (C 1s) region of the XPS spectrum of diamond is shown in Figure 1.1b.

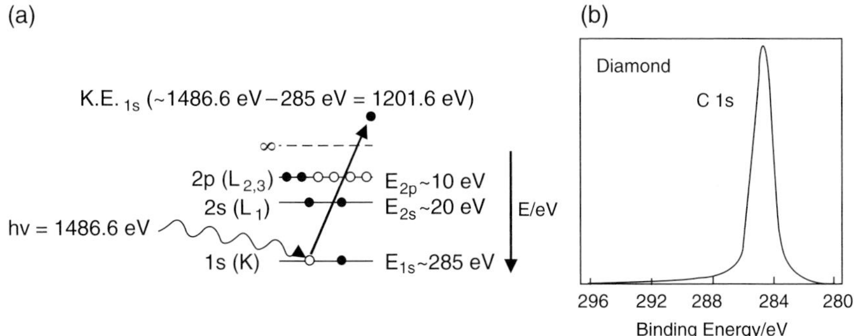


Figure 1.1 (a) Schematic of the XPS process, showing photoionisation of an atom by the ejection of a 1s electron from an atom of carbon; (b) The C 1s region of the XPS spectrum of diamond. [Note: the arrangement of the electrons in diamond differs from that shown in (a) because, in diamond, the 2s and 2p orbitals form a set of four equivalent $sp^3$ hybrid orbitals resulting in a single peak in the XPS spectrum.]

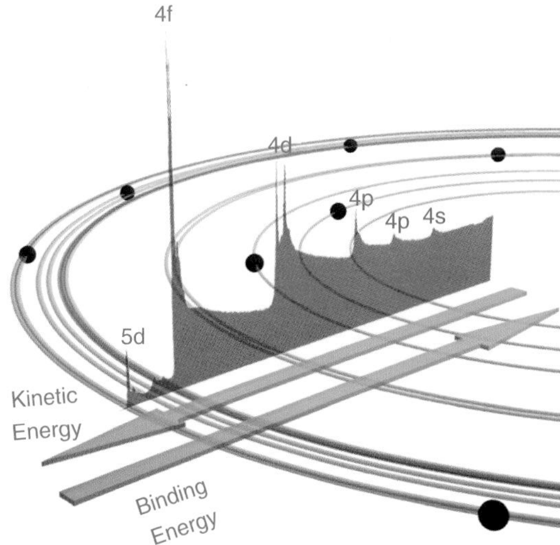


Figure 1.2 Photo electron spectrum of lead showing the manner in which electrons escaping from the solid can contribute to discrete peaks or suffer energy loss and contribute to the background. The spectrum is superimposed on a schematic of the electronic structure of lead to illustrate how each orbital gives rise to photoelectron lines.

The photoelectron spectrum will reproduce the electronic structure of an element quite accurately as all electrons with a binding energy less than the photon energy will feature in the spectrum. This is illustrated in Figure 1.2 where the XPS spectrum of lead is superimposed on a representation of the electron orbitals (represented as circular orbits in Figure 1.2).

Those electrons which are excited and escape without energy loss contribute to the characteristic peaks in the spectrum; those which undergo inelastic scattering and suffer energy loss contribute to the background of the spectrum. Once a photoelectron has been emitted, the ionised atom must relax in some way. This can be achieved by the emission of an X-ray photon, known as X-ray fluorescence (XRF). The other possibility is the ejection of an Auger electron. Auger peaks may be seen in the XPS spectrum of tin shown in Figure 1.3. As will be seen later, the position of the Auger peaks in the XPS spectrum can provide valuable chemical information.

Figure 1.3 shows the XPS spectrum of Sn with the peaks labelled according to the above notation and illustrating the splitting observed in the peaks due to electrons in 3p and 3d orbitals while splitting in the 4d and 4p peaks is too small to be observed.

XPS provides sensitive elemental analysis of a surface, but its great advantage is that it can provide a chemical state analysis as well. This is because the binding energy of the core electrons in an atom is affected by the chemical environment of the atom. Figure 1.4 illustrates this with a C 1s spectrum of poly(ethylene terephthalate) (PET). The spectrum consists of three major peaks; one due to the carbon atoms in the aromatic ring (carbon with no oxygen attached), one due to

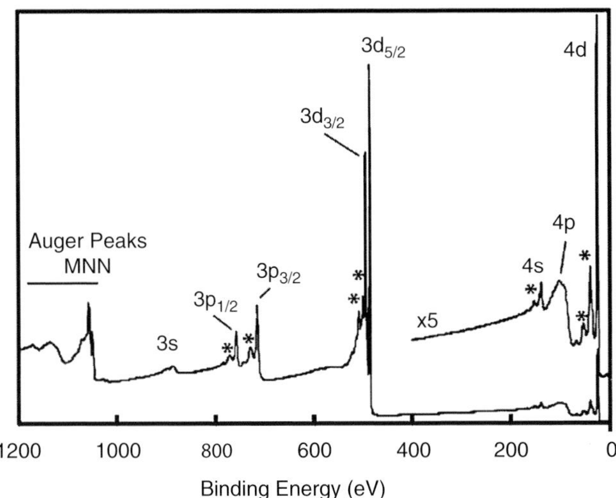


Figure 1.3 Survey spectrum from Sn showing the XPS transitions accessible using Al Kα radiation. The features marked with an asterisk are electron energy-loss features due to plasmon excitation (see Section 3.3.8).

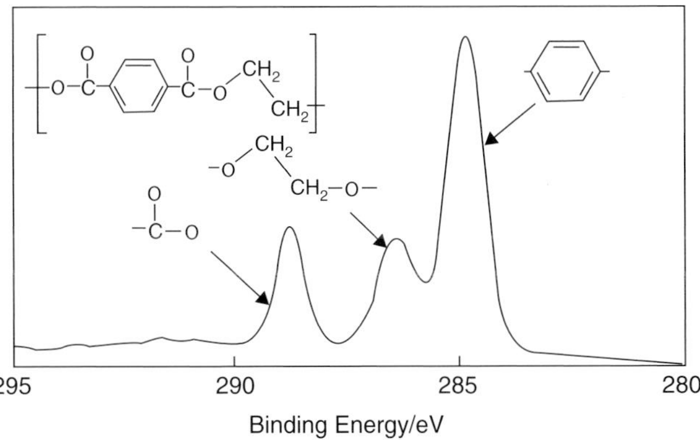


Figure 1.4 The C 1s region of the XPS spectrum of poly (ethylene terephthalate) (PET).

the carbon atoms in the ethylene group (carbon with one oxygen atom attached), and one due to the carbon atoms in the ester group (carbon with two oxygen atoms attached). It should be noted that the oxygen atoms in the two different chemical environments also show a chemical shift in their 1s spectrum.

Table 1.3 shows examples of C 1s binding energies for a range of compound types, many more may be found in, for example, the National Institute of Standards and Technology (NIST) data base (https://srdata.nist.gov/xps/Default.aspx).

Table 1.3 C 1s binding energies for a range of compound types.

| Chemical state | Example | C 1s Binding energy/eV |
|---|---|---|
| $sp^2$ carbon | Graphite, graphene, aromatic polymer, e.g. polystyrene | 284.4 |
| C-C | Polyethylene | 284.8 |
| $sp^3$ carbon | Diamond | 284.8 |
| C-O | Poly(ethylene terephthalate) (PET) | 286 |
| C-Cl | Poly(vinyl chloride) (PVC) | 287 |
| C=O | Poly(ethyl ketone) (PEK) | 288 |
| O=C-O | Poly(methyl methacrylate) (PMMA) | 288.5 |
| $CF_2$ | Polytetrafluoroethylene (PTFE) | 292 |
| $CO_3^{2-}$ | Metal carbonate | 288–290 |
| $C^{n-}$ | Metal carbide | 283–284 |
| $CF_3$ | $CF_4$ plasma treated polymers | 294 |

## 1.4 Auger Electron Spectroscopy (AES)

The processes of Auger electron emission and X-ray emission are illustrated in Figure 1.5. A high-energy incident electron causes the emission of an electron from an atom, (a K electron in this example). The incident electron loses energy and is scattered within the sample. The scattered incident electron and the emitted K electron contribute to the background signal in an Auger spectrum. The emission of the K electron from the atom leaves a 'hole' in the electronic structure which means that the ionised atom is in an excited state and will relax to a lower energy state in one of two ways.

1) An electron from a higher level, an L electron in this example, fills the hole causing an X-ray photon to be emitted. This is the basis of electron probe microanalysis (EPMA), carried out in many electron microscopes by either energy dispersive X-ray spectroscopy (EDX) or wavelength dispersive X-ray spectroscopy (WDX) spectrometers.
2) The core hole may be filled by an electron from a higher level, the $L_{2,3}$ level in Figure 1.5. In accordance with the principle of the conservation of energy, another electron is ejected from the atom, e.g. an $L_1$ electron in the schematic of Figure 1.5. This electron is termed the $KL_1L_{2,3}$ Auger electron.

With increasing atomic number, the relative yield of Auger electrons decreases, see Figure 1.6, while the yield of X-ray photons increases. Implicit in this observation is the fact that AES, in common with XPS, has good sensitivity for light elements.

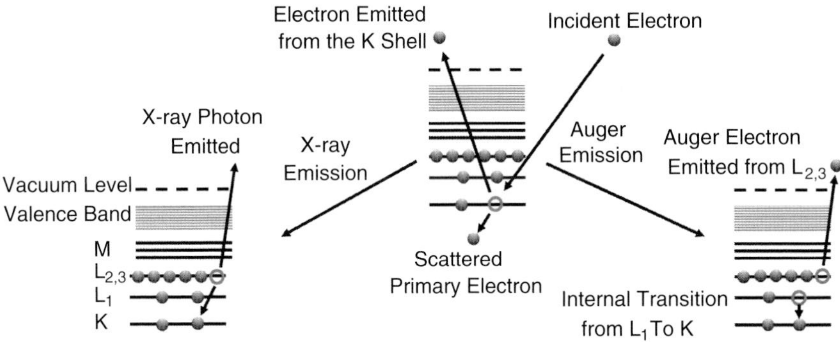


Figure 1.5 The competing processes of X-ray photon emission and Auger electron emission as a result of the ejection of a core level electron caused by a high-energy, incident electron.

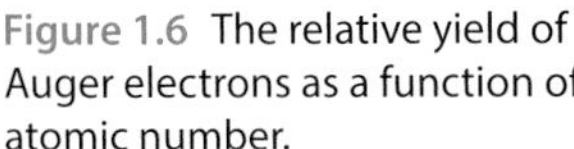
Figure 1.6 The relative yield of Auger electrons as a function of atomic number.

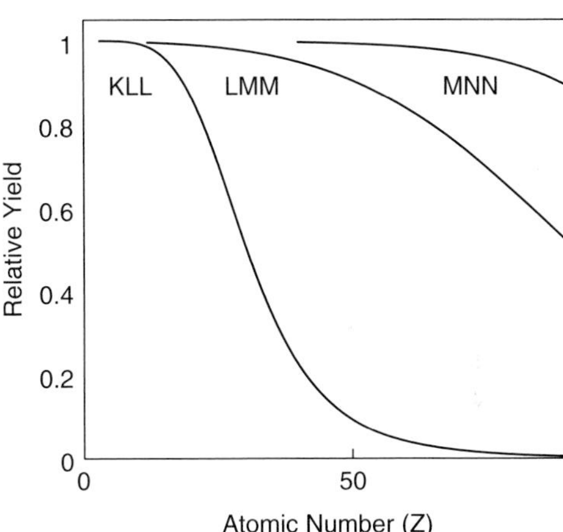


Abbreviated forms of the Auger notation are often used, for example:

- *KLL* is used, omitting the subscripts, to refer to the whole group of *KLL* emissions from silicon.
- *NVV* is a general term used to refer to Auger emissions in which an electron is removed from the N orbital, replaced by an electron from the valence shell, causing a second valence electron to be emitted.
- *CVV* is a general term to indicate the involvement of one electron from a core level and two from the valence band. The broadness of the valence band is convoluted within the final Auger distribution leading to very broad peaks.
- *VVV* is a general term to indicate Auger emissions involving three valence electrons.

In most cases, it is the *CCC* Auger transitions (in which all three electrons originate from core or core-like orbitals) which provide chemical information in AES.

The kinetic energy of a $KL_{2,3}L_{2,3}$ Auger electron is approximately equal to the difference between the energy of the core hole and the energy levels of the two outer electrons, $E(L_{2,3})$ (The term $L_{2,3}$ is used in this case because, for light elements, $L_2$ and $L_3$ cannot be resolved):

$$\mathrm{E}\left(\mathrm{KL}_{2,3}\mathrm{L}_{2,3}\right) \approx \mathrm{E}(\mathrm{K}) - \mathrm{E}\left(\mathrm{L}_{2,3}\right) - \mathrm{E}\left(\mathrm{L}_{2,3}\right)$$

This equation does not take into account the interaction energies between the core holes ($L_{2,3}$ and $L_{2,3}$) in the final atomic state nor the inter- and extra-relaxation energies which come about as a result of the additional core screening needed. Thus, the calculation of the energy of Auger electron transitions is much more complex than the simple model outlined above, but there is a satisfactory empirical approach which considers the energies of the atomic levels involved and those of the next element in the periodic table.

Following this empirical approach, the Auger electron energy of transition $KL_1L_{2,3}$ for an atom of atomic number Z is written:

$$E\left(KL_1L_{2,3}, Z\right) = \mathrm{E}\left(K, Z\right) - \tfrac{1}{2}\left[E\left[L_1, Z\right] + E\left[L_1, Z+1\right]\right]$$
$$-\tfrac{1}{2}\left[E\left[L_{2,3}, Z\right] + E\left[L_{2,3}, Z+1\right]\right]$$

Clearly for the $KL_{2,3}L_{2,3}$ transition the second and third terms of the above equation are identical, and the expression can be simplified to:

$$E\left(KL_{2,3}L_{2,3}, Z\right) = E\left(K, Z\right) - \left[E\left[L_{2,3}, Z\right] + E\left[L_{2,3}, Z+1\right]\right]$$

It is the kinetic energy of this Auger electron, ($KL_{2,3}L_{2,3}$), that is the characteristic material quantity irrespective of the primary beam composition (i.e. electrons, X-rays, ions) or its energy. For this reason, Auger spectra are always plotted on a kinetic energy scale.

Figure 1.7 shows the Auger spectrum of nickel. The upper (direct) spectrum is simply the signal in counts or counts per second plotted against electron kinetic energy. The lower spectrum is the differential spectrum. Many analysts prefer the use of the differential form of the Auger spectrum.

AES does not have the chemical specificity that XPS exhibits. It is, however, frequently used to distinguish metals and their oxides or nitrides, for example. Figure 1.8 shows the Auger spectrum from an aluminium foil which has a native oxide at its surface. Here, the metallic peak is clearly separated from the peak attributed to oxidised aluminium. This spectrum also exhibits a plasmon peak which is associated with the metallic aluminium and results from the electrons emitted from the metal atoms losing energy by exciting valence-band

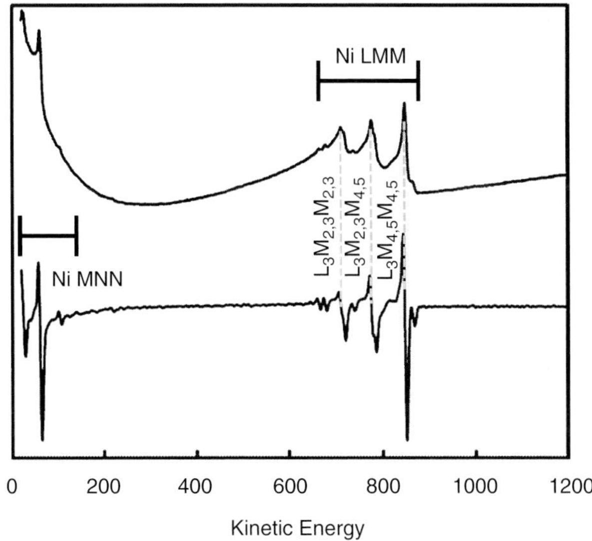


Figure 1.7 The Auger spectrum of nickel, the upper curve is the direct spectrum while the lower one is the differential spectrum.

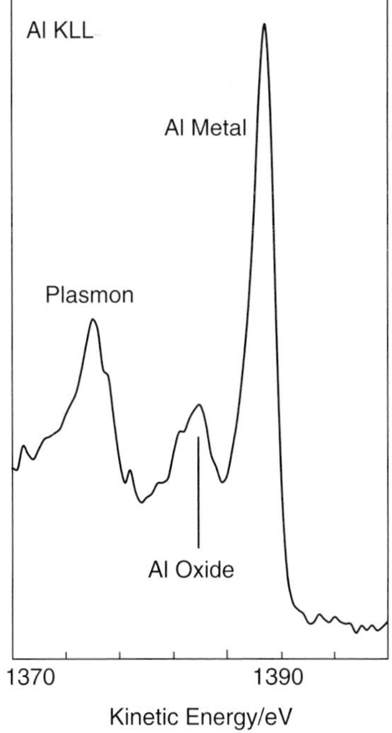


Figure 1.8 Part of the Auger spectrum of aluminium foil which has a thin native oxide at its surface. The spectrum shows that the metal and its oxide can be clearly resolved in the Auger spectrum. Also shown in the spectrum is a plasmon peak.

Table 1.4 Modes of analysis available with scanning Auger microscopy (SAM).

| Mode of analysis | Scanned | Fixed |
|---|---|---|
| Point analysis | E | x, y |
| Line scan | x | E, y |
| Chemical map | x, y | E |

electrons and generating collective oscillations. (A plasmon is the excitation of valence band electrons in a solid in which collective oscillations are generated, this will be described in Section 3.3.8).

## 1.5 Scanning Auger Microscopy

The use of a finely focused electron beam for AES enables us to achieve surface analysis at a high spatial resolution, in a manner analogous to EPMA in the scanning electron microscope (SEM). By combining an electron spectrometer with an (UHV) SEM, it becomes possible to carry out Scanning Auger Microscopy (SAM). In the scanning Auger microscope, various modes of operation are available; the variable quantities being the position of the electron probe on the sample (x and y) and the setting of the electron energy analyser (E) corresponding to the energy of emitted electrons to be analysed. Table 1.4. summarises the possibilities.

When an instrument having multiple detectors is used, it is usually possible to collect data at a number of fixed energies simultaneously, such as peak and background energies which are required for topography correction as described below. Alternatively, appropriate regions of the spectrum may be collected at each point in a linescan or map allowing maps to be constructed following processing of the spectral data, a form of retrospective analysis.

As the Auger electron yield is very sensitive to electron take-off angle, an image of Auger electron intensities will invariably reflect the surface topography of the sample, possibly more strongly than the chemical variations. This problem is overcome by recording a background (B) as well as the Auger peak (P) map. However, a simple subtraction of the background counts from the peak intensity (P-B) is not sufficient, as shown in Figure 1.9a. The use of a simple algorithm such as (P-B)/B, allows correction for the effects of surface topography. This may be seen in Figure 1.9b, where variation in intensity resulting from the topography of the material has been largely suppressed and only chemical information remains.

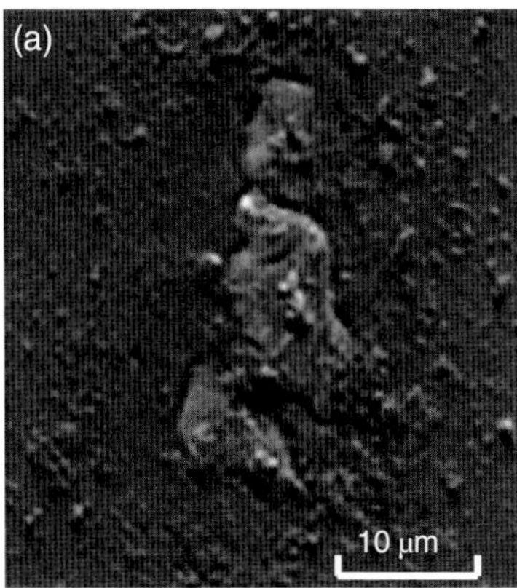


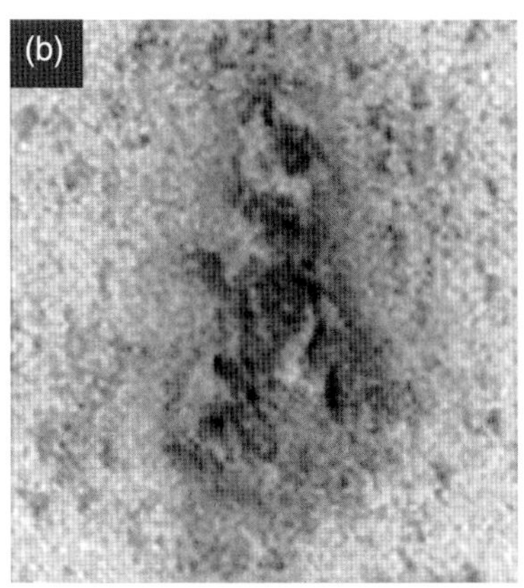


Figure 1.9 Scanning Auger microscopy (SAM) of the O KLL peak from an inclusion group in an aluminium alloy (a) peak-background map (P-B), (b) correction for topographic effects using (P-B)/B algorithm.

## 1.6 The Depth of Analysis in Electron Spectroscopy

The depth of analysis in both XPS and AES is dependent upon the ability of the electrons to escape from the solid without losing energy. The depth from which this can occur is small compared with the depth to which the incident radiation can penetrate.

Electrons passing through a solid material may undergo either elastic or inelastic collisions with the atoms in a solid. Electrons which have undergone inelastic collisions lose some of their kinetic energy and, therefore, no longer contribute to the characteristic peak in the XPS spectrum. The intensity of electrons (I) emitted from all depths greater than d in a direction normal to the surface is given by the Beer-Lambert relationship:

$$I = I_0 \exp(-d/\lambda)$$

Where $I_0$ is the intensity from an infinitely thick, uniform substrate. The quantity $\lambda$ will be defined below but it is typically of the order of a few nanometres for photoelectrons having a kinetic energy in the region of 1 keV and depends upon the physical properties of the material through which it is travelling.

For electrons emitted at an angle $\theta$ to the surface normal, this expression becomes:

$$I = I_0 \exp\left[-d\cos(\theta)/\lambda\right]$$

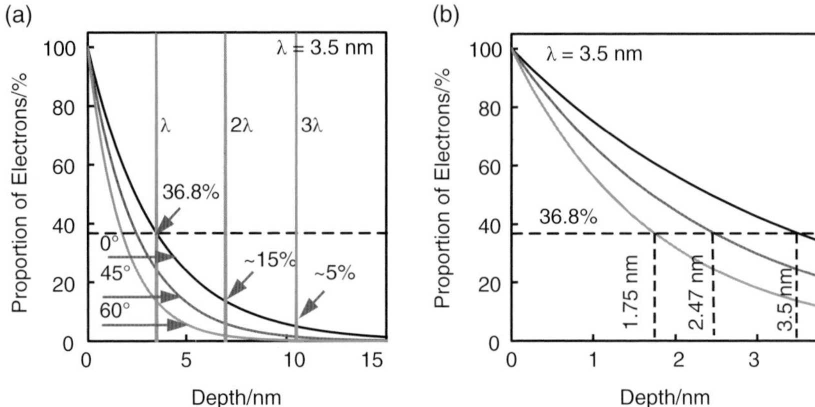


Figure 1.10 (a) The proportion of electrons emitted from greater than the indicated depth when the value of λ is 3.5 nm. Curves are given for emission angles of 0°, 45°, and 60°; (b) Curves as in (a) but with an expanded depth scale. Indicated on this plot are the depths from which 63.2% of the electrons are emitted (36.8% are emitted from greater depths).

Using the appropriate analysis of the above equations, it can be shown that by considering electrons that emerge with an emission angle of 0° (90° to the sample surface), 63.2% of the signal in electron spectroscopy is expected to come from a depth of $<\lambda$, 86.5% from a depth of $<2\lambda$, and 95% from a depth of $<3\lambda$. This is illustrated in Figure 1.10a which shows the proportion of electrons emitted from depths less that λ, 2λ and 3λ. An arbitrary value of 3.5 nm has been chosen for λ in this figure. This is the mechanism by which the signal from the vast majority of the atoms present in the sample is filtered out (stated as a requirement for a surface analysis technique in Section 1.1).

The Beer-Lambert equation can be manipulated in a variety of ways to provide information about overlayer thickness and to provide a non-destructive depth profile (that is without removing material by mechanical, chemical, or ion-milling methods).

It is tempting to assume that the value of the inelastic mean free path (IMFP)[3] of the electrons is the value that should be used for λ. Unfortunately, this does not lead to the correct values for depth or thickness being obtained. The reason for this is largely due to the presence of elastic scattering in the sample. Following an elastic scattering event, the direction in which the electron is travelling changes but its kinetic energy does not change significantly and so it will contribute to the XPS peak. This leads to the intensity of the detected signal at an emission angle, θ, being greater than expected from the simple model. The reason for this is illustrated in Figure 1.11. This shows two electrons, A and B, emitted from atoms at different depths within the sample end

3 The inelastic mean free path of an electron is defined in ISO 18115 to be the average distance that an electron with a given energy travels between successive inelastic collisions.

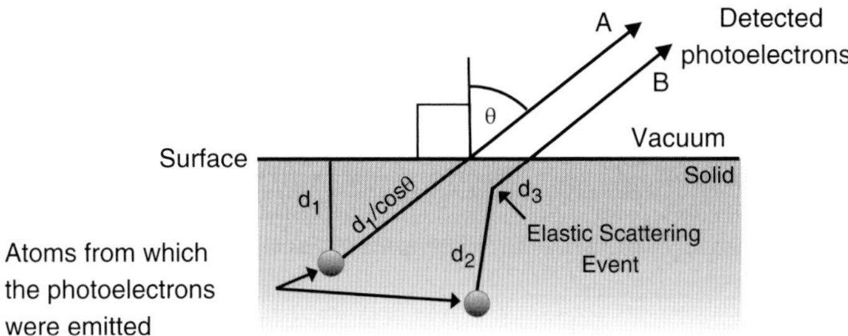


Figure 1.11 An illustration of the way in which elastic scattering can enhance the signal observed at large emission angles.

emerging from the surface of the sample at the same emission angle (θ). The atom that emits electron A is at a distance $d_1$ below the surface and this electron is emitted without being scattered after travelling a distance of $d_1/\cos(\theta)$. Electron B travels a distance $d_2$ towards the surface before being scattered and then leaves the surface after travelling a distance $d_3$. In this example, the two electrons have travelled the same distance inside the sample ($d_1/\cos(\theta) = d_2 + d_3$) but electron B begins its journey at a depth below the surface greater than $d_1$. The effect of this phenomenon increases with increasing emission angle. A commonly stated 'rule of thumb' when using angular data to calculate layer thicknesses is to avoid the use of data collected at emission angles greater than 60°. This will be discussed further in section 4.2.4.

The attenuation of the electron current as it passes through a solid broadly follows the Beer-Lambert equation but it is necessary to determine the value of λ for each case. Typically, the value of λ is about 20% less than the IMFP and is called the 'attenuation length'.[4] The term 'effective attenuation length'[5] is also used in XPS and Auger.

For XPS, an empirical description of λ was proposed by Seah and Dench (1979) of the National Physical Laboratory and is given below:

$$\lambda = \frac{538a_A}{E_A^2} + 0.41a_A\left(a_A E_A\right)^{0.5}$$

where $E_A$ is the energy of the electron in eV, $a_A^3$ is the volume of the atom in $nm^3$ and λ is in nm. This equation fitted the experimental data with a standard devia-

4 The attenuation length of an electron is defined in ISO 18115 to be the quantity λ in the expression $\Delta x/\lambda$ for the fraction of a parallel beam of specified particles or radiation removed in passing through a thin layer $\Delta x$ of a substance in the limit as $\Delta x$ approaches zero, where $\Delta x$ is measured in the direction of the beam.

5 The effective attenuation length of an electron is defined in ISO 18115 to be the parameter which, when introduced in place of the inelastic mean free path into an expression derived for AES and XPS on the assumption that elastic scattering effects are negligible for a given quantitative application, will correct that expression for elastic scattering effects.

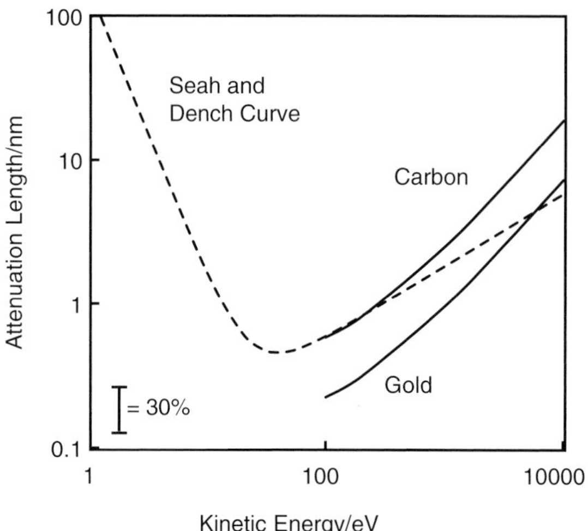


Figure 1.12 The Seah and Dench 'universal curve' and curves for carbon and gold calculated from the more recent expression.

tion of about 30%. The curve is shown in Figure 1.12. More recently, Seah has published an expression which fits the data for elements to within about 8%. This provides the analyst with a very straightforward route to the estimation of attenuation length and is preferred to the original Seah and Dench equation, above.

$$\lambda = \left(0.65 + 0.007E^{0.93}\right) / Z^{0.38}$$

The original Seah and Dench curve and examples of curves from the more recent expression are shown in Figure 1.12.

Values for λ and IFMP may be derived from optical spectroscopy, reflection electron energy loss spectroscopy (REELS) as well as from XPS and Auger spectroscopy if suitable, characterised layered materials are available. Various databases exist from which values of IMFP and attenuation length can be obtained.

It follows from the above that the depth of analysis in electron spectroscopy depends upon:

- The material being analysed.
- The kinetic energy of the electrons being measured.
- The energy of the incident X-ray photons (because this will determine the kinetic energy of the electrons emitted from a given orbital).
- The emission angle.

## 1.7 Comparison of XPS and AES/SAM

Although it is difficult to make a comparison of techniques before they are described and discussed in detail, it is pertinent at this point to outline the strengths and weaknesses of each to provide background information.

XPS was also known by the acronym ESCA[6] (electron spectroscopy for chemical analysis). It is this *chemical* specificity which is the major strength of XPS as an analytical technique, one for which it has become deservedly popular. By this, we mean the ability to define not only the elements present in the analysis but also their chemical state. In the case of iron, for instance, the spectra of $Fe^0$, $Fe^{2+}$, and $Fe^{3+}$ are all slightly different and, to the expert eye, are easily distinguishable. However, such information is attainable in XPS only at the expense of spatial resolution, and XPS is usually regarded as an area integrating technique. Small area XPS (SAXPS) is available on modern instruments and, when operating in this mode, a *spectroscopic* spatial resolution of <10 µm is often possible. Some modern instruments offer imaging XPS, and such imaging may have a spatial resolution of <3 µm. Set beside a spatial resolution of ~10 nm which can be achieved on the latest commercial Auger microprobes, it becomes clear that the XPS is not the way to proceed for surface analysis at very high lateral resolution, but the advantages of the levels of information available from an XPS analysis (the ease with which a quantitative analysis can be achieved, its greater applicability to insulators, and the ready availability of chemical state information) will often offset this.

In addition to the chemical state information referred to above, XPS spectra can be quantified in a very straightforward manner and meaningful comparisons can be made between samples of a similar type. Quantification of Auger data is rather more complex, and the accuracy obtained is generally not so good. Because of the complementary nature of the two methods and the ease with which Auger and photoelectron analyses can be made on the same instrument, the two methods have come to be regarded as the most important methods of surface analysis in the context of materials science. All manufacturers of electron spectrometers offer both XPS and AES options for their systems.

## 1.8 The Availability of Surface Analytical Equipment

The capital cost of an XPS/AES/SAM spectrometer is high when compared with most electron microscopes, it is of the order of £0.5–£1.0 M (at the time of writing) for a comprehensive system. This, allied to the fairly steep learning curve that the newcomer must ascend before confidence in the technique is obtained, has led to the development of laboratories offering surface analysis as a service facility. Such laboratories may be found throughout the world. They are often associated with universities but the balance between academic and industrial work varies greatly. The use of service facilities presents a very

---

6 The term 'ESCA' was first coined by the Swedish Nobel Laureate Kai Siegbahn in 1967. He used this term to emphasise the fact that chemical, not just elemental, information is available from the technique. The term is now deprecated and the term 'XPS' will be used throughout the remainder of this book.

Figure 1.13 A modern high-performance XPS system.

attractive proposition to inexperienced users in that expert advice is always on hand to ensure the efficient use of instrument time; a factor that is of paramount importance as the daily charge for the use of such a facility can exceed £2000. It is not unusual for analysts to 'cut their teeth' on the field of surface analysis in such a way. Once the need within their company (and their own personal expertise) has been established, a surface analysis system can be specified for their own particular needs.

Although originally the exclusive preserve of research laboratories and academic institutions, surface analysis facilities are now frequently to be found in industrial trouble-shooting and quality assurance roles. As the techniques find wider applications, so the market grows and manufacturers are very willing to continue developing their spectrometers. Increasing automation allows XPS instruments to be used by non-specialists giving the instruments a wider appeal within both industry and academia.

XPS stations can be found on synchrotron light sources. These have two major advantages over stand-alone, laboratory instruments. The first is the very high intensity of the X-ray beam. This shortens acquisition times and can allow the sample to be analysed at a higher pressure leading to more realistic conditions for catalytic, electrochemical, or biological studies. The second major advantage is that the energy of the incident X-ray beam is tuneable, meaning that the analysis depth can be controlled over a wide range. Although the use of a synchrotron-based system provides many unique analytical capabilities, there are limited opportunities to use such an instrument.

A modern, commercial electron spectrometer is shown in Figure 1.13.

# 2

# Electron Spectrometer Design

## 2.1 Introduction

The design and construction of electron spectrometers is a complex undertaking and will usually be left to one of the handful of specialist manufacturers worldwide, although many users specify minor modifications to suit their own requirements. The various modules necessary for analysis by electron spectroscopy are (in addition to a sample); a source of the primary beam (either X-rays or electrons), an electron energy analyser and detection system, all contained within a vacuum chamber, and a data system, which is an integral part of the instrument.

## 2.2 The Vacuum System

Most commercial spectrometers are based on vacuum systems designed to operate in the ultra-high vacuum (UHV) range of $10^{-8}$ to $10^{-10}$ mbar,[1] and most XPS and AES experiments are carried out in this pressure range. The reason for this is twofold:

1) The analytical signal of low energy electrons is easily scattered by the residual gas molecules and, unless their concentration is kept to an acceptable level, the total spectral intensity will decrease, whilst the noise present within the spectrum will increase.
2) More importantly, the UHV environment is necessary because of the surface sensitivity of the techniques themselves. At $10^{-6}$ mbar, it is possible for a monolayer of gas to be adsorbed onto a solid surface in about one second. A typical analysis takes much more than one second to complete.

---

1 Traditionally, pressures in UHV equipment are measured in mbar. 1 mbar is equivalent to 100 Pascals (Pa). 1 Pa is equal to 1 $Nm^{-2}$.

*An Introduction to Surface Analysis by XPS and AES*, Second Edition.
John F. Watts and John Wolstenholme.

Some sample types are not suited to a UHV environment as the vacuum can alter the structure or composition of the surface. To accommodate this type of sample, there are now commercially available instruments in which the sample is placed in a relatively high-pressure environment (~25 mbar). Care must be taken with this type of instrument to ensure that the path length of the electrons in the high-pressure region is minimised and that the pressure within the spectrometer and source regions is sufficiently low. Further details concerning XPS measurements in these 'near ambient' conditions, near ambient pressure XPS (NAPXPS) will be given in section 2.6.

The manner in which such a vacuum is established will depend on customer and manufacturer preferences. The chambers and associated piping will invariably be made of stainless steel and joints will usually be made by using flanges, equipped with knife-edges, which are tightened onto copper gaskets (a system generally referred to as *Conflat* following the designation by Varian Associates who own the trademark but which are now supplied by a number of manufacturers). For some joints single or double O-ring seals are used. The O-rings are typically made from a synthetic rubber and fluoropolymer elastomer material.

UHV conditions are frequently obtained in a modern electron spectrometer using ion pumps or turbomolecular pumps; the latter are the better choice if there is a requirement to pump large quantities of noble gases. Diffusion pumps, which were very popular some time ago, have now disappeared from modern commercial instruments. Whichever type of pump is chosen, it is common to use a titanium sublimation pump to assist the prime pumping and to achieve the desired vacuum level. All UHV systems need baking from time to time to remove adsorbed layers from the chamber walls. The baking temperature is dictated by the analytical options fitted to the spectrometer but is usually in the range 100–160 °C for routine use.

In a stand-alone electron spectrometer, the vacuum system consists of at least two vacuum chambers; the analysis chamber and the sample introduction chamber. The latter is required so that samples can be loaded into the instrument without exposing the analysis chamber to atmospheric pressure. Ideally, the introduction chamber should have a relatively small volume so that it can be pumped quickly to a high vacuum before the sample is transferred to the analysis chamber. The introduction chamber on some instruments also serves as a preparation chamber, housing preparation facilities (e.g. a fracture stage, a high-pressure gas cell, etc.). Some instruments consist of three chambers; an introduction chamber, a preparation chamber and an analysis chamber. A three-chamber system has the advantage that the preparation chamber may be maintained at UHV conditions.

A further requirement of the analysis chamber is that it supports the analytical components with the required rigidity, geometry, and alignment. As an example, a schematic diagram of a modern scanning Auger microscope is shown in

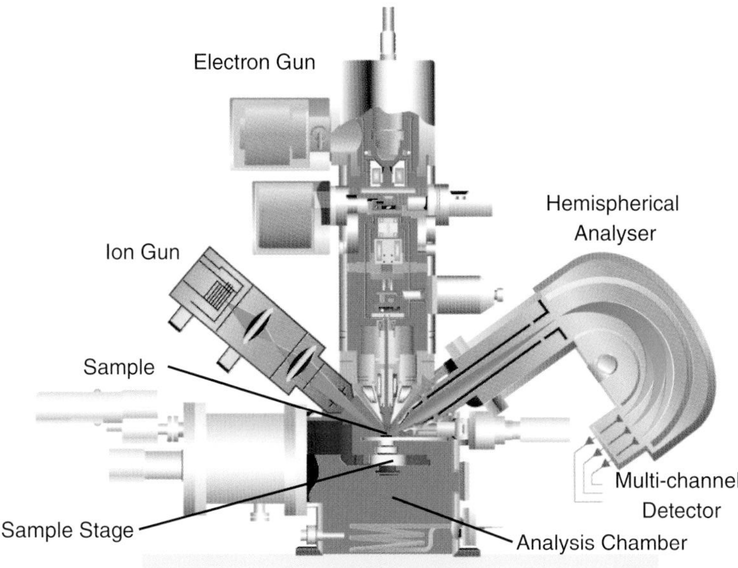


Figure 2.1 Cross section of a scanning Auger microscope showing the positions of the analytical components of the instrument.

Figure 2.1, the major components of the instrument are labelled and will be described in detail later.

During analysis, the sample must be attached to a sample stage or sample manipulator the purpose of which is to hold the sample in the correct position and orientation for the analysis. Most high-end electron spectrometers have a five-axis stage, having x, y, z, tilt, and rotate motions. The x, y and z motions allow the analyst to position the sample such that the point or area to be analysed is positioned in the correct position relative to the analytical components of the instrument (i.e. the analysis position). Sample tilt is required so that the analyst can select the angle of incidence or emission that is most appropriate for the analysis. Sample rotation is usually used during depth profiling to optimise depth resolution (see Chapter 4). The motions of the stage are motorised and controlled via the data system of the instrument. Sample heating and cooling facilities are often fitted to the sample stage.

The trajectory of low-energy electrons is strongly influenced by the Earth's magnetic field. Consequently, some form of magnetic screening is required around the sample and electron energy analyser. There are two approaches to this problem. The most elegant solution is to fabricate the entire analysis chamber from a material with high magnetic permeability ($\mu$-metal). An acceptable alternative is to fabricate shielding panels, either as sleeving within the instrument or as a bolt-on outer shroud. The methodology depends on the manufacturer. In addition, compensation coils may be arranged around the analyser and transfer lens to mitigate the effect of external magnetic fields.

## 2.3 X-ray Sources for XPS

X-rays are generated by bombarding an anode material with high-energy electrons. The electrons are emitted from a thermal source which may either be an electrically heated tungsten filament or a lanthanum hexaboride ($LaB_6$) emitter. The $LaB_6$ emitter is used for focusing X-ray monochromators because of its higher current density (brightness).

The efficiency of X-ray emission from the anode is determined by the electron energy, relative to the X-ray photon energy. For example, the Al K$\alpha$ photon (energy 1486.6 eV) flux from an aluminium anode increases by a factor of more than 5 if the electron energy is increased from 4 to 10 keV. At a given energy, the photon flux from an X-ray anode is proportional to the electron current striking the anode. The maximum anode current is determined by the efficiency with which the heat, generated at the anode, can be dissipated. For this reason, X-ray anodes are usually water-cooled.

The properties of the X-ray beam play an important role in determining both the quality of the XPS spectrum and the nature of the information it provides. The five most important properties are:

1) **Energy**: The energy of the X-ray photons will dictate the maximum binding energy that can be observed in the spectrum. The energy also determines the analysis depth.
2) **Line width**: The line width directly affects the width of the photoelectron peak in the XPS spectrum. No XPS peak can be narrower than the width of the X-ray line.
3) **Purity**: X-ray anodes produce a series of X-ray lines, each at a different energy and each having a different intensity. These lines will give rise to a set of overlapping XPS spectra.
4) **Spot size and shape**: This is important if the X-ray beam is used to define the area of analysis. Under these conditions the spot size and shape will dictate the lateral resolution of the analysis.
5) **Flux**: The number of photoelectrons detected per second is directly related to the number of X-ray photons falling on the sample within the area being analysed. If the flux is too small the analysis will take a long time to reach the required signal-to-noise ratio. It may be impossible to achieve the required ratio in some cases. If the flux is too great, then the detector will saturate leading to non-linearity and poor quantification.

Of these, the flux is the easiest parameter for the analyst to control. It is done by controlling the electron current falling on the X-ray anode. There is, of course, a maximum flux permitted for a given anode. Exceeding this maximum will cause the anode material to melt.

The X-ray energy is determined by the material selected for the anode material, to be discussed in the next section.

If an X-ray monochromator is used, both the line width and X-ray purity are optimised for analysis.

Some monochromators are designed to focus the beam of X-rays onto the sample. The size and shape of the X-ray spot on the sample is then determined by the size and shape of the electron beam striking the X-ray anode. This too is under the control of the analyst.

It is clear from above that the first choice of the analyst would be to use an X-ray source fitted with a monochromator. This would not have been true in the past because the sensitivity of the instruments was such that spectra produced using the monochromator would take too long to acquire. Now that there have been marked improvements in both spectrometer sensitivity and monochromator design, the sensitivity has improved to the point where monochromators are used unless there is a compelling reason not to use one. Indeed, many modern, high-end instruments have only a monochromated X-ray source, some do not even have the option to fit a non-monochromated source. By far the most popular monochromated X-ray source for XPS is one based on Al Kα radiation.

### 2.3.1 Choice of X-ray Anode

The choice of anode material for XPS determines the energy of the X-ray transition generated. Ideally, it should be of sufficiently high photon energy to excite an intense photoelectron peak from all elements and it must possess a natural X-ray line width that will not excessively broaden peaks in the resultant spectrum. The most popular anode materials (for non-monochromated X-ray sources) are aluminium and magnesium. These are usually supplied in a single X-ray gun with a twin anode configuration providing Al Kα or Mg Kα photons of energy 1486.6 and 1253.6 eV respectively. This is possible because, unlike anodes for X-ray diffraction (XRD), it is the anode and not the filament which is at a high potential (for XRD the filament is at a high negative potential and the anode at ground; for XPS the filament is at or near ground and the anode at a high positive potential of 10–15 kV).

Such twin anode assemblies are useful as they provide a modest depth profiling capability – the difference in the analysis depth of organic materials of the carbon 1s electron is about 1 nm greater for electrons excited by Al Kα, compared with Mg Kα. More importantly, they provide the ability to differentiate between Auger and photoelectron transitions when the two overlap in one radiation. XPS peaks will change to a position 233 eV higher on a kinetic energy scale on switching from Mg Kα to Al Kα whereas the energy of Auger transitions remains constant. On a binding energy scale, of course, the reverse is true as shown in Figure 2.2.

The use of elements other than magnesium and aluminium is becoming more popular. Table 2.1 shows a list of possible anode materials, along with their energy, full width at half maximum intensity (FWHM)[2] and the heaviest

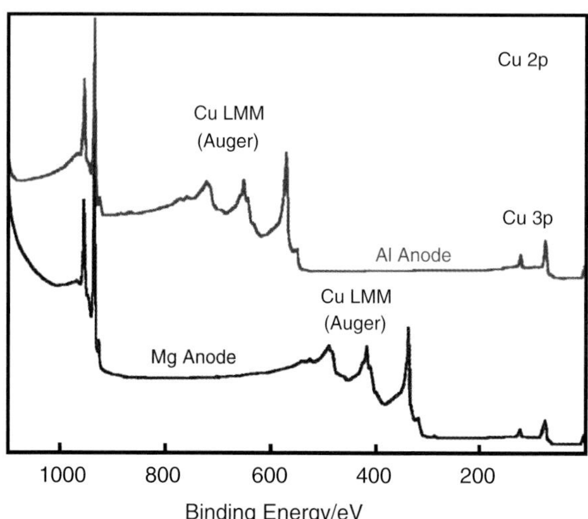


Figure 2.2 Comparison of XPS spectra recorded from copper using AlKα (upper) and MgKα (lower) radiation. Note that on a binding energy scale, the XPS peaks remain at constant values but the X-ray induced Auger electron (X-AES) transitions move by 233 eV on switching between the two sources.

Table 2.1 Possible anode materials for XPS.

| Element | Line | Energy /eV | FWHM /eV | 'Deepest' 1s[a] |
|---|---|---|---|---|
| Mg | $K\alpha_{1,2}$ | 1253.6 | 0.7 | Na |
| Al | $K\alpha_{1,2}$ | 1486.6 | 0.9 | Mg |
| Si | Kα | 1739.6 | 1.0 | Al |
| Zr | Lα | 2042.4 | 1.7 | Si |
| Ag | Lα | 2984.4 | 2.6 | Cl |
| Ti | Kα | 4510.9 | 2.0 | Ca |
| Cr | Kα | 5417.0 | 2.1 | Ti |
| Cu | $K\alpha_1$ | 8047.8 | 2.2 | Ni |
| Ga | Kα | 9251.7 | 2.6 | Zn |

[a] The heaviest element for which the 1s photoelectron can be measured using the indicated X-ray anode.

2 The values quoted for the FWHM represent the natural width of the stated line. When used, a monochromator will reduce this line width considerably (see Section 2.3.2).

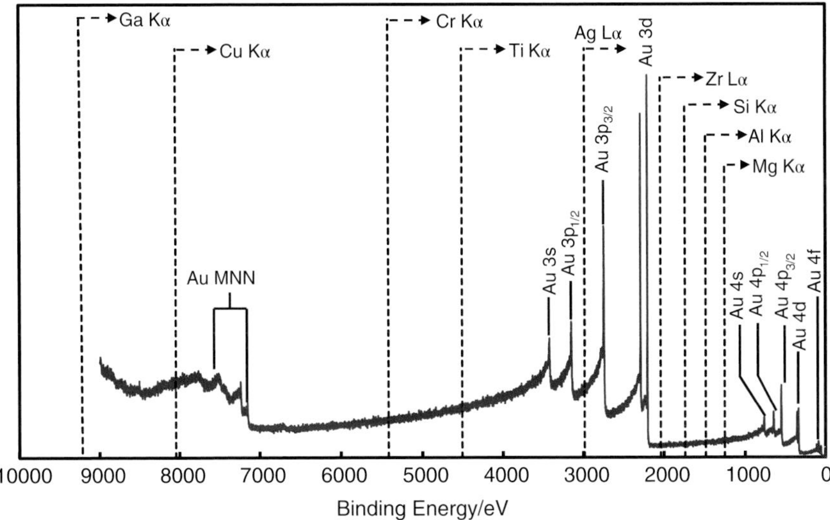


Figure 2.3 XPS spectrum of gold acquired using a gallium anode. The maximum binding energy that can be measured using the anode materials of Table 2.1 is also indicated on the spectrum.

element for which the 1s photoelectron can be detected in the XPS spectrum using the stated anode material. At the time of writing, there are commercially available high-energy XPS systems based on Ag Lα, Cr K$\alpha_1$ and Ga K$\alpha_1$.

In a twin anode arrangement, any two of these anode materials (excluding gallium) can be used in any combination. The extent to which the binding energy scale is extended using higher energy X-ray sources is indicated in Figure 2.3. The X-ray excited photoelectron spectrum of gold is presented up to a binding energy of 9000 eV. In conventional, Al Kα excited XPS only the 4f, 4p and 4s orbitals are present in the spectrum. The extension of the binding energy scale with an alternative photon source is indicated with dotted lines. For example, XPS using a Ag Lα source enables Au 3d and Au 3p3/2 to be accessed and so on.

Gallium in an X-ray anode is used in its liquid state and this affects the mechanical design of the instrument. The gallium is used in the form of a metal jet, the liquid gallium is ejected from a nozzle in vacuum. Electrons hit the jet and X-rays are generated. The gallium is contained in a closed loop in the X-ray tube and the return line of gallium is cooled. The anode is not cooled directly.

Gallium is an ideal material for this type of source because it has a very low melting point (29.76 °C) and a high boiling point (2400 °C) which means that its vapour pressure remains compatible with a UHV environment over a wide temperature range, so a high-power electron beam may be used. The liquid in the jet constantly exposes a fresh clean surface to the electron beam and, because there is no substrate, there is no possibility of ghost peaks appearing (this can happen on a conventional source when the anode material gets worn exposing the substrate).

Although it is not possible to include gallium as an anode in a conventional twin-anode source, in principle, it would be possible to add a metal to the gallium to form an alloy. The added metal will produce its own X-ray lines and the required line selected using a suitable monochromator. This type of arrangement is not currently available commercially.

When using anodes which generate high-energy photons, maximum benefit is derived if the analyser is capable of measuring electron kinetic energies approaching the energy of the photon. Unless this is the case, the low binding energy peaks would not be detected.

There are two advantages of higher energy anodes.

1) As already mentioned, energy levels not available using the more conventional X-ray anodes become accessible.
2) Because the use of higher energy photon sources results in higher kinetic energy of the ejected photoelectrons, the analysis depth is greater. It is therefore possible to gain information concerning the variation of composition with depth merely by changing the X-ray source. Figure 2.4 shows the dependence of the attenuation length of Si 2p, Si 1s, Ag 3d and Ag 2p electrons upon the kinetic energy of these electrons and, therefore upon the anode material used in the X-ray source.

Figure 2.4 shows the relationship between the kinetic energy of the photoelectrons and the photon energy for photoelectrons emitted from silicon (Si 2p and Si 1s) and silver (Ag 3d and Ag 2p). Obviously, if the photon energy is less than the binding energy for a given transition then the electron will not be emitted. Figure 2.4 also shows how the ionisation cross section of Si 2p, Si 1s (Figure 2.4a), Ag 3d and Ag 2p (Figure 2.4b) decrease with the kinetic

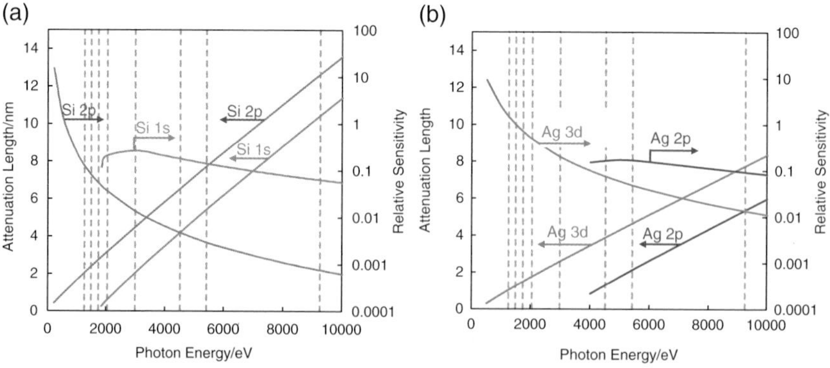


Figure 2.4 The attenuation length and the ionisation cross section of (a) Si 2p, Si 1s, Ag 3d and (b) Ag 2p as a function of incident photon energy. The vertical dashed lines show the photon energies available from the anode materials listed in Table 2.1.

energy of the photoelectrons. This, of course, will have an effect upon the intensity of the peaks in the photoelectron spectrum. In this Figure, the photoelectron emission cross section, coupled with the analysis volume have been used to show the relative photoelectron emission current normalised to the Ag 3d transition from Al Kα radiation. The curves do not take account of instrument transmission function and they assume that the photon flux density is the same for all measurements.

Using a Ga Kα X-ray source, at an energy of 9.252 keV, the attenuation length of electrons forming XPS peaks of low binding energy is approximately 10 times greater than those emitted using an Al Kα source (see Figure 2.4) and so the information obtained from the sample is more representative of the bulk material than the surface.

The generation of high-energy X-rays requires the use of high-energy electrons, typically, electrons must have a kinetic energy in the region of 70 keV as they strike the anode.

If XPS measurements are performed on a beamline of a synchrotron, the analyst is not constrained to use only those X-ray energies mentioned above. The X-ray energy may be tuned over a wide range to a value which is optimised for the measurements being made.

Conducting XPS measurements using high-energy photons is often referred to as hard X-ray photoelectron spectroscopy (HAXPS or HAXPES). HAXPS used to be the preserve of measurements made using synchrotron radiation but there are now laboratory-based instruments which provide an HAXPS capability. X-rays having a photon energy greater than 5 keV are generally considered to be hard X-rays in this context.

### 2.3.2 X-ray Monochromators

The purpose of an X-ray monochromator is to produce a narrow X-ray line by using diffraction in a crystal lattice. Figure 2.5 illustrates the process. X-rays strike the parallel crystal planes at an angle θ and are reflected at the same angle. The distance travelled by the X-rays depends upon the crystal plane at which they are reflected. Figure 2.5a shows two adjacent crystal planes with X-rays being reflected from each. If the distance between the planes is d then the difference in the path length is 2d sin θ. If this distance is equal to an integral number of wavelengths then the X-rays interfere constructively, if not destructive interference takes place. This gives rise to the well-known Bragg equation:

$$n\lambda = 2\mathrm{d}\sin\theta$$

In which

$$\begin{aligned} \mathrm{n} &= \text{the diffraction order} \\ \lambda &= \text{the X-ray wavelength} \\ \mathrm{d} &= \text{the crystal lattice spacing} \\ \theta &= \text{Bragg angle (angle of diffraction)} \end{aligned}$$

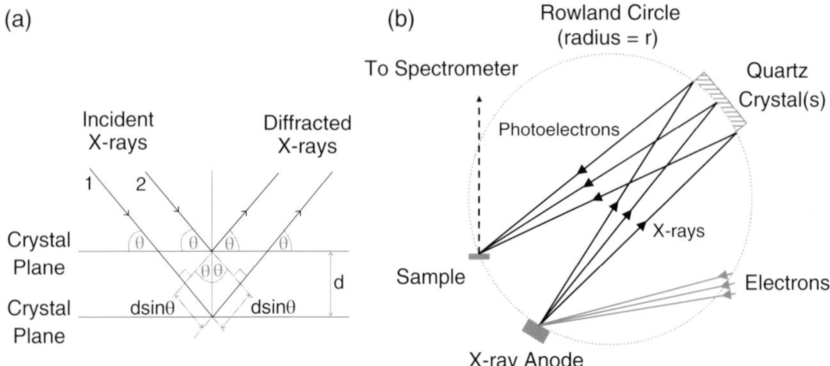


Figure 2.5 (a) Diffraction of X-rays at a quartz crystal; (b) the positioning of the sample and monochromator components on the Rowland circle on a focusing monochromator.

At present, all commercially available X-ray monochromators used to produce Al Kα radiation for XPS employ a quartz crystal (usually the $[10\bar{1}0]$ crystal face) as the diffraction lattice. The position of the crystal relative to the anode and sample is important. These components must be arranged on the 'Rowland circle' (of radius, r), as illustrated in Figure 2.5b.

If the crystal surface is planar with its centre touching the Rowland circle, the X-ray beam at the sample will be in the form of a line extending out of the plane of the diagram in Figure 2.5b. To produce an X-ray beam focused at the sample, the crystal must be curved. The curvature is obtained by pressing and bonding a thin wafer of quartz onto a precisely machined glass substrate. The radius of curvature of the crystal must be twice that of the Rowland circle. This arrangement is known as the 'Johann geometry'. If the crystal, curved with a radius of 2r, is further ground to a radius of r, this becomes the 'Johansson geometry'. The latter provides more precise focusing but is more difficult to manufacture and so the Johann geometry is usually used. Some monochromators are fitted with multiple crystals to increase the flux of X-rays at the sample. The crystals must be accurately aligned with each other and with the spectrometer and so adjusters are usually provided so that adjustment can be made from outside the vacuum system.

If the quartz crystal is curved in such a way that it focuses the X-ray beam as well as causing it to be diffracted, the size of the X-ray spot on the surface of the sample is approximately equal to that of the electron spot on the anode. Thus, by varying the focus of electron source, the analyst can vary the analysis area. If the electron beam is scanned or rastered over the anode it follows that the X-ray beam is scanned over the surface of the sample. There are commercially available instruments that use this method as a basis for XPS imaging, see Section 2.9.

Most commonly, the monochromator on an XPS instrument is used for Al Kα radiation. However, other materials and other diffraction orders have been used. For example, there are commercially available XPS instruments which have a dual anode X-ray monochromator (aluminium and silver). This means that the user can switch between these anodes as required by the analysis. It should be noted that the wavelength of the Ag Lα line is not exactly half of that of the Al Kα line which means that the geometry of the monochromator must be adjusted slightly each time a change from one radiation to the other is made. In addition, being the result of a second-order diffraction, the intensity of the Ag Lα beam is considerably lower than that of the Al Kα beam, so data acquisition times are typically correspondingly longer. In principle, Ti Kα and Cr Kβ radiation may be used in a similar manner with the Ti Kα using the third-order diffraction and Cr Kβ using the fourth. There is a commercially available instrument that includes a chromium anode but a quartz crystal is not used. This instrument uses a Cr/Al anode and two crystals so that the analyst can select the required radiation by means of a simple shutter mechanism. An alternative crystal material is used for a gallium X-ray source.

Quartz is a convenient material because it is relatively inert, compatible with UHV conditions, it can be bent and/or ground into the correct shape and its lattice spacing provides a convenient diffraction angle for Al Kα radiation.

There are several reasons for using an X-ray monochromator on an XPS spectrometer:

1) The primary reason is the reduction in X-ray line width, for example, from 1.0 eV to approximately 0.25 eV for Al Kα, and from 2.6 to 1.2 eV for Ag Lα. Narrower X-ray line width results in narrower XPS peaks and consequently better chemical specificity.
2) Unwanted portions of the X-ray spectrum (i.e. satellite peaks and the bremsstrahlung continuum) are also removed.
3) For maximum sensitivity, a twin anode X-ray source is usually positioned as close to the sample as possible. The sample is therefore exposed to the radiant heat from the source region which could damage or alter the surface of delicate samples. When a monochromator is used, this heat source is remote from the sample and thermally induced damage is avoided.
4) It is possible to focus X-rays into a small spot using the monochromator. This means that small area XPS (SAX) can be conducted with high sensitivity.
5) Use of a focusing monochromator means that only the area of the sample being analysed is exposed to X-rays. Thus, a number of samples may be

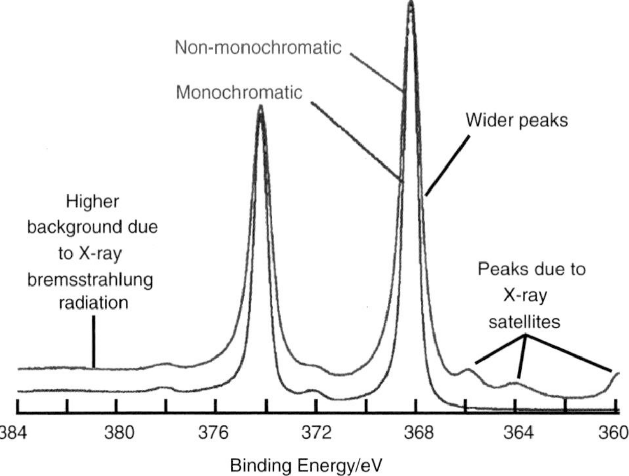


Figure 2.6 A comparison of the Ag 3d spectra acquired with monochromatic and non-monochromatic X-rays. (The spectra are normalised to the maximum peak intensity).

loaded into the spectrometer without the risk of X-rays damaging samples whilst they await analysis. Similarly, multi-point analysis can be performed on the same delicate sample.

Some of these advantages are illustrated in Figure 2.6. This shows the XPS spectrum of Ag 3d acquired using monochromatic and non-monochromatic X-rays. The analyser was set to the same conditions for each spectrum. There is a clear difference in the peak width, the background is higher using the non-monochromatic X-rays and X-ray satellites are clearly visible when the non-monochromatic source is used.

Using a focusing X-ray monochromator, illustrated in Figure 2.5, it is possible to produce a SAX analysis and this now forms the basis of commercial instruments with a spatial resolution of < 15 μm. This is one route to SAX, the other commercially available method, electron-optical aperturing, is discussed later in this chapter.

### 2.3.3 Synchrotron Sources

X-rays derived from a synchrotron source are advantageous for XPS for two major reasons. First, the photon energy may be fine-tuned over a wide range of energy, allowing control over the analysis depth. Second is the fact that the intensity of the X-ray beam is very high allowing shorter data acquisition times and analysis at higher ambient pressures. Pressures in the region of 10 mbar in the vicinity of the sample may be used allowing the sample to be analysed in more realistic conditions.

## 2.4 The Electron Gun for AES

Since 1969 (the year in which Auger electron spectrometers became commercially available), AES has become a well-accepted analytical method for the provision of surface analyses at high spatial resolution. The intervening years have seen the electron guns improve from 500 μm resolution of the converted oscilloscope gun of the late 1960s to the < 10 nm resolution which is typical of top-of-the range Auger microprobes of today. In between these two extremes are the 5 μm and 100 nm guns which are typical of electron guns in use with instruments on which Auger is not the prime analytical technique (multi-technique instruments).

The critical components of the electron gun are the electron source and the electron column which includes the lens assemblies for beam focusing and shaping along with the components required for scanning the beam. Nowadays, the combination of a field emission source of electrons and electromagnetic lenses provides Auger instruments whose electron spot size can be less than 10 nm.

### 2.4.1 Electron Sources

To be useful as an electron source for AES, a source should have the following properties:

1) **Stability**: The current emitted from the source should be highly stable over long periods. Although spectra can be obtained in a few minutes, if the spectrometer is to be used for depth profiling then stability over many hours is essential.
2) **Brightness**: High-emission currents from a small emitted area are required if the eventual spot size at the sample is to be small.
3) **Mono-energetic**: The focal length of electromagnetic and electrostatic lenses is dependent upon the energy of the electrons. This means that the optimum focusing conditions can only occur for electrons having a very small range of kinetic energy. A wide energy spread will therefore result in a large spot size at the sample.
4) **Longevity**: Under normal operating conditions, the emitter must not need to be replaced for many hundreds of hours. Replacement of emitters requires that the vacuum is broken and the instrument baked before it can be used. This operation means that the instrument cannot be used for a period of two or three days.

Several different types of source have been used in commercial instruments; the more important of these will now be described.

#### 2.4.1.1 Thermionic Emitter

The simplest form of thermionic source is a tungsten wire fabricated in the form of a hairpin, or 'V' shape. When an electric current is passed through the wire, the temperature rises giving electrons sufficient energy to overcome the work function and to be released into free space. The work function is the energy required for an electron to escape from a solid surface; for tungsten, this is about 4.5 eV. By reducing the work function, the number of electrons emitted per unit area per unit solid angle can be increased, thereby increasing the so-called *brightness* of the source. Shaping the emitter into the form of a hairpin (or 'V') reduces the area of the filament which contributes to the electron beam and therefore minimises the spot size. This reduction is partly due to geometrical reasons and partly because the electrical voltage drop across the emitter area is minimised thus reducing the energy spread.

Simple thermionic emitters are inexpensive and robust but they lack brightness and so it is difficult, using this type of source, to attain spot sizes for Auger analysis below about 200 nm.

#### 2.4.1.2 Lanthanum Hexaboride Emitter

A widely-used material for high-brightness sources is single-crystal lanthanum hexaboride ($LaB_6$). The source consists of a small, indirectly heated crystal of $LaB_6$. The crystal is cylindrical in shape and conical at one end. The tip of the cone is ground to form a flat surface of about 15 μm in diameter and it is important that the flat surface exposes the <100> crystal face.

This material has a much lower work function than tungsten (2.6 eV compared with 4.5 eV), which means that it has a high-emission current density even at a much lower temperature. The operating temperature for a $LaB_6$ emitter is about 1800 K whereas 2300 K is typical for a tungsten emitter.

The lower operating temperature and the fact that there is no voltage drop across the emitting surface mean that the energy spread in the electron beam is small compared with that from a tungsten emitter.

For stable operation, this type of emitter requires a better vacuum than the tungsten emitter.

#### 2.4.1.3 Cold Field Emitter

A type of emitter which has found use in Auger microscopy is that based on the field emission source. The operating principle of a field emission source is not to give the electrons sufficient energy to jump the work function barrier (as in the thermionic process) but to reduce the magnitude of the barrier itself, both in height, (marginally), but more importantly in width. It is the latter factor which leads to improved electron emission, as electrons from near the Fermi level can penetrate the barrier by quantum tunnelling and thus escape from the emitter with no loss in energy.

Another characteristic of the field emission source is its narrow energy distribution, as there are no electrons above the Fermi level and those below it have a rapidly decreasing probability of escape. In practical terms, this feature means a smaller spot size because chromatic aberrations within the electron lenses are small.

Usually, field emission is achieved by the application of a very large electrostatic field between the emitter, which itself must be in the form of a needle with a tip radius of approximately 50 nm, and an extraction electrode. The small area of emission from the tip into a small solid angle provides a high brightness compared with thermionic sources, although the total current will be somewhat lower. The emitter material usually employed is a tungsten single crystal.

A cold-field emitter in a practical Auger system will provide very small spot sizes, and therefore excellent spatial resolution but suffers from a lack of long-term stability unless operated in an extremely good vacuum ($<10^{-10}$ mbar). The presence of residual gases close to the emitter will have two destabilising effects:

1) The gases will adsorb on the cold surface contaminating it and affecting its emission current. The emitting area of this type of source is so small that even a single adsorbed molecule on the emitting surface will have a noticeable effect on the emission current.

   Adsorbed gases can be removed by 'flashing' the tip. This process involves heating the tip to a temperature near its melting point, causing the adsorbed gases to be removed and, if flashing is undertaken in an electrostatic field, a reshaping of the tip. The reshaping is usually beneficial but occasionally it is such that the tip becomes inoperable and must be replaced. Following flashing, emission from the tip will be unstable for a period of about an hour.
2) Some of the high-energy electrons in the beam will collide with the residual gas molecules forming positive ions. These ions are then accelerated in the electrostatic field and collide with the tip, causing sputtering (removal of tungsten atoms from the surface). This affects the radius of the tip and reduces the efficiency of the emission process. The very low pressure in the vicinity of the tip minimises this effect but does not eliminate it.

#### 2.4.1.4 Hot Field Emitter

This type of electron emitter (also known as a Schottky field emitter) has become very popular in recent years because its brightness is high and stability good. Although this type of source is often called a 'field emission source', it would be more accurately called a 'field-assisted thermionic emitter'. The field does not cause electrons to be emitted by quantum tunnelling, it causes the effective work function of the emitting surface to be reduced, see Figure 2.7.

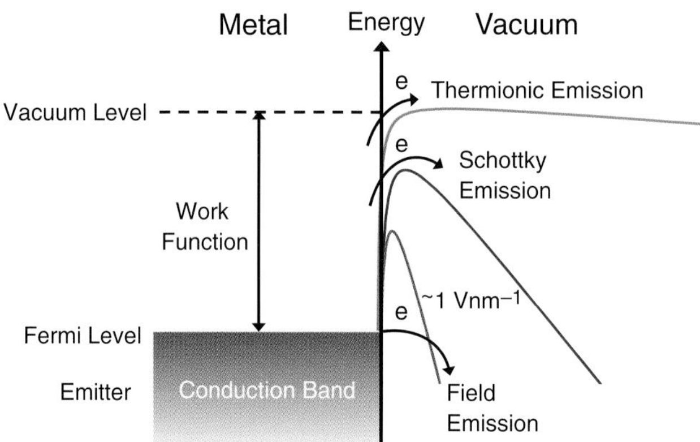


Figure 2.7 The emission process from the three types of emitter described here.

The source consists of a single crystal tungsten wire coated with the semiconductor material zirconium oxide ($ZrO_2$). The function of the $ZrO_2$ is:

- to lower the work function of the emitter and increase the emission current
- to provide a self-cleaning surface
- to provide a self-healing surface.

Again, the emitting surface is small (about 20 nm in diameter) and so little demagnification is required to achieve the small spot sizes required in high-resolution scanning Auger microscopy. The tip is heated to about 1800 K in a large electrostatic field. The combination of the high temperature and the field causes electrons to be emitted.

The vacuum requirements for this type of gun are much less stringent than is the case for the cold field emitter, although it is usual to provide additional pumping at the source when this type of emitter is used.

#### 2.4.1.5 Comparison of Electron Emitters for AES

The emission process for each type of emitter is illustrated in Figure 2.7. The effect of an increasing electric field in the vicinity of the emitter upon the effective work function of the material can be seen. When the field is very high (~1 $Vnm^{-1}$), electrons do not have to have sufficient energy to surmount the barrier but can leave the emitter by quantum tunnelling.

Table 2.2 compares the important characteristics of the types of emitter considered here.

The stringent vacuum requirements and the relatively poor stability mean that the cold field emitter is rarely used for AES. For medium- and high-performance Auger instruments, either $LaB_6$ or Schottky field emitters are used in commercially available instruments.

Table 2.2 Comparison of electron emitters.

| | Thermionic | $LaB_6$ | Cold field emitter | Schottky emitter |
|---|---|---|---|---|
| Work Function (eV) | 4.5 | 2.7 | 4.5 | 2.95 |
| Brightness ($Acm^{-2}srad^{-1}$) | $<10^5$ | $\sim 10^6$ | $10^7$ to $10^9$ | $>10^8$ |
| Current into a 10 nm spot | 1 pA | 10 pA | 10 nA | 5 nA |
| Maximum Beam Current | 1 μA | 1 μA | 20 nA | 200 nA |
| Minimum Energy Spread | 1.5 eV | 0.8 eV | 0.3 eV | 0.6 eV |
| Operating Temperature (K) | 2700 | 2000 | 300 | 1800 |
| Short Term Stability | <1% | <1% | >5% | <1% |
| Long Term Stability | High | High | >10%/hr | <1%/hr |
| Vacuum Required (mbar) | $<10^{-4}$ | $<10^{-6}$ | $<10^{-10}$ | $<10^{-8}$ |
| Typical Lifetime (h) | <200 | ~1000 | >2000 | >2000 |
| Relative Cost | Low | Medium | High | High |

### 2.4.2 The Electron Column

The purpose of the electron column is to produce a small spot of electrons on the sample and to scan that spot over the surface of the sample to produce images and line scans. In many cases, the lenses in the column will form a demagnified image of the source at the sample. The column usually consists of no less than two lenses, the condenser lens and the objective lens. To achieve optimum performance, the beam must be accurately aligned with the axis of the column, this requires sets of alignment coils in a high-performance gun or mechanical adjusters in a more basic design.

Beam current is dependent upon the electron emission current, the setting of the condenser lens and the size of the aperture placed in the electron path after the condenser lens.

A high-performance gun will scan the beam using electromagnetic coils whilst a gun fitted to a multi-technique instrument is likely to have electrostatic scan plates.

### 2.4.3 Spot Size

The spot size attainable with a particular electron gun is a function of the primary beam current. For example, the smallest spot size obtainable on a scanning Auger microscope with electromagnetic lenses and $LaB_6$ filament is about 20 nm at 0.1 nA but increases to 100 nm at 10 nA. The intensity of Auger electrons emitted depends on the sample current and at 0.1 nA spectrum acquisition will be a lengthy process, but at 10 nA the current of Auger electrons

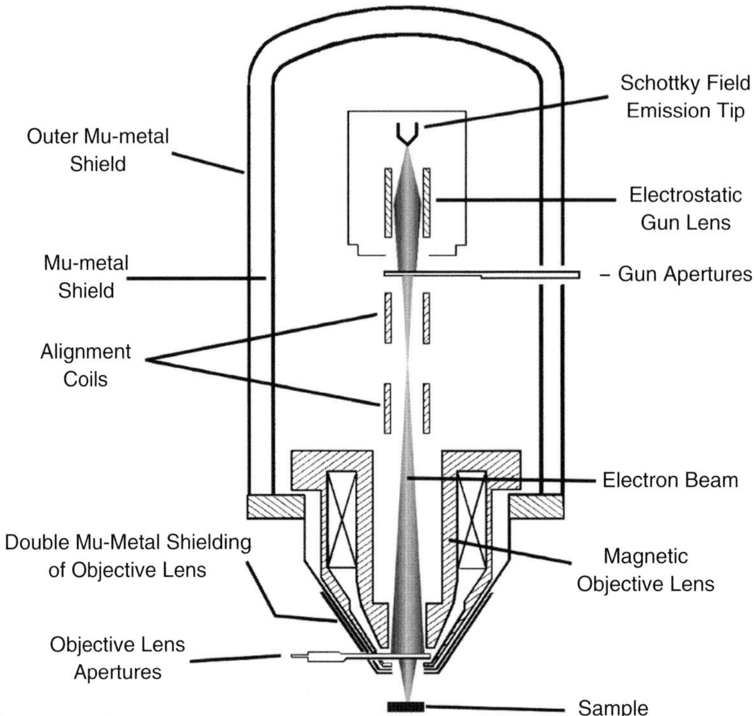


Figure 2.8 Schematic of an ultra-high vacuum (UHV) electron gun for Auger electron spectroscopy (AES).

will be increased to the point where analysis becomes practical. A satisfactory compromise must be reached between spatial resolution and spectral intensity. A field emission source in a similar column will give somewhat better resolution but at a much-improved current owing to its superior brightness. Such electron guns are the preserve of high-resolution scanning Auger microscopes; such a gun is illustrated in Figure 2.8.

When SAM is required on a multi-technique XPS instrument, the best configuration probably includes a high-brightness, hot field emission gun with electrostatic lenses. Such an assembly will provide reliable routine operation with the minimum of attention and an analytical spot size of $< 100\,\text{nm}$.

A modern, high-end scanning Auger microscope (shown schematically in Figure 2.1) equipped with a thermally assisted field emission source and a hemispherical sector electron energy analyser is shown in Figure 2.9.

The similarities compared with a scanning electron microscope are readily appreciated. This configuration has the advantage that additional analytical facilities, such as X-ray analysis (see Section 5.6), backscattered electron

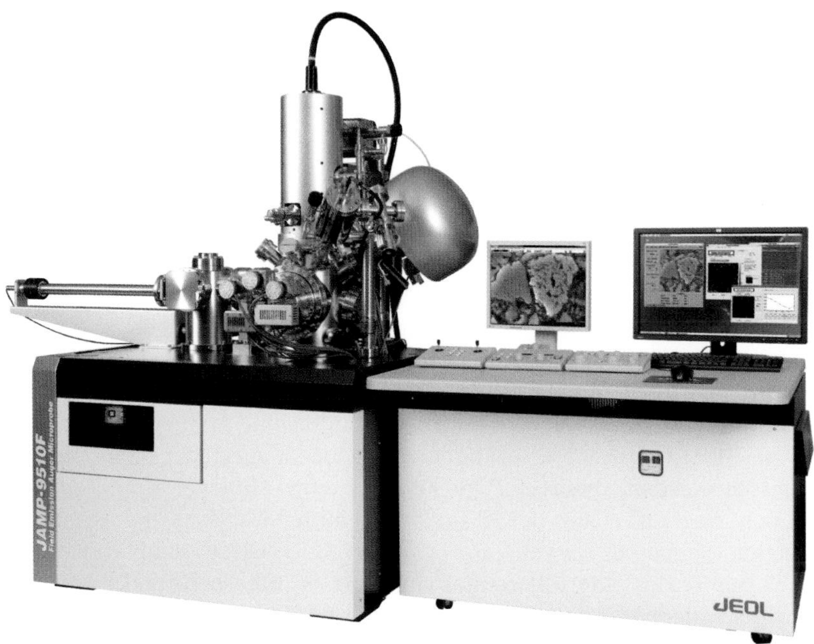


Figure 2.9 A modern high-performance scanning Auger microscope.

imaging and XPS (by the addition of a twin anode X-ray source) are options that are readily implemented. In particular, the provision of energy dispersive X-ray analysis on a scanning Auger microscope enables simultaneous analysis of the bulk and surface regions of the sample.

## 2.5 Analysers for Electron Spectroscopy

There are two types of electron energy analyser in general use for XPS and AES; the cylindrical mirror analyser (CMA) and the hemispherical sector analyser (HSA).

The development of these two analysers reflects the requirements of AES and XPS at the inception of these techniques. The primary requirement for Auger spectroscopy was that of high-sensitivity (analyser transmission), the intrinsic resolution (the contribution of analyser broadening to the resultant spectrum) being of minor importance. The need for high sensitivity led to the development of the CMA. For XPS, on the other hand, it is spectral resolution that is the cornerstone of the technique and this led to the development of the HSA as a design of analyser with sufficiently good resolution. The addition of a transfer lens to the HSA and multi-channel detection increases its sensitivity to the point where both high transmission and high resolution are possible, and this type of analyser may now be used with excellent results for both XPS and AES. The nature of the CMA does not permit the addition of a transfer lens.

### 2.5.1 The Cylindrical Mirror Analyser

The CMA consists of two concentric cylinders as illustrated in Figure 2.10. The inner cylinder is held at earth potential whilst the outer is ramped at a negative potential. An electron gun is often mounted coaxially within the analyser. A certain proportion of the Auger electrons emitted will pass through the defining aperture in the inner cylinder, and, depending on the potential applied to the outer cylinder, electrons of the desired energy will pass through the detector aperture and be re-focused at the electron detector. Thus, an energy spectrum – the direct energy spectrum – can be built up by merely scanning the potential on the outer cylinder to produce a spectrum of intensity (in counts per second) versus electron kinetic energy. This spectrum will contain not only Auger electrons but all the other emitted electrons, the Auger peaks being superimposed, as weak features, on an intense background signal.

For this reason, the differential spectrum is often recorded rather than the direct energy spectrum. In the past, this used to be achieved by applying a small AC modulation to the analyser and comparing the output at the detector with this standard AC signal by means of a phase sensitive detector (lock-in-amplifier). The resultant signal is displayed as the differential spectrum. In modern Auger spectrometers, however, the differential spectrum, if required, is calculated within the data system. A comparison of direct

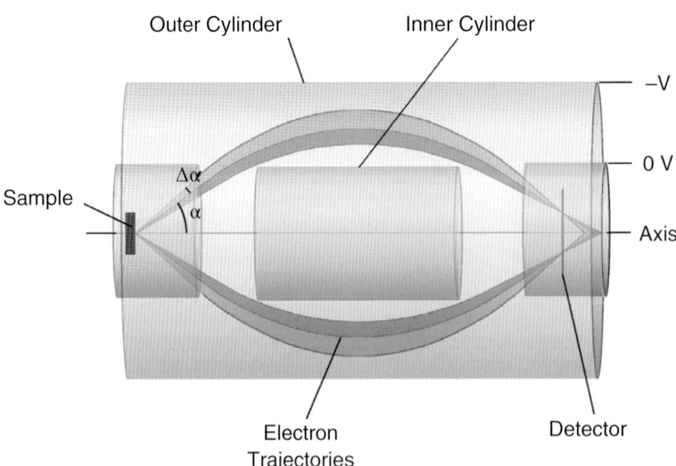


Figure 2.10 Schematic diagram of the cylindrical mirror analyser (CMA).

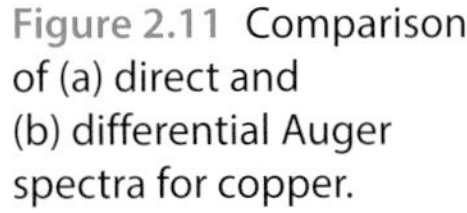

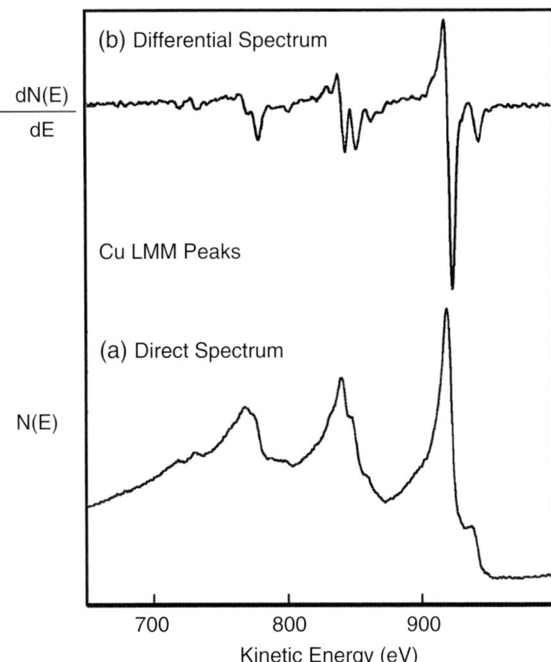


Figure 2.11 Comparison of (a) direct and (b) differential Auger spectra for copper.

(pulse counted) and differential Auger electron spectra, from a copper foil, is presented in Figure 2.11.

Whilst the CMA can provide good sensitivity for AES, it suffers from a number of disadvantages which make it totally unsuitable for XPS:

- The optimum resolution obtainable with this type of analyser is not good enough to provide the chemical state information which is available from XPS.
- The energy calibration of the analyser is dependent upon the position of the sample surface along the axis of the analyser.
- The area from which electrons can be collected is very small.

In an attempt to overcome these disadvantages, a double-pass CMA was developed with limited success. All modern, commercial XPS instruments are now equipped with an HSA.

A device has been developed by one manufacturer to improve upon the resolution of the CMA, Figure 2.12. When this device is in use, a positive potential is applied to the sample, situated below a grounded electrode which has an aperture centred on the axis of the CMA. This means that the electrons are retarded as they travel between the sample and the grounded electrode. The kinetic energy of the electrons is therefore reduced as they pass through the CMA and so the resolution is improved, as follows.

The resolution of the CMA for electrons of kinetic energy $E_1$,

$$\Delta E_0 = RE_1 \qquad (R = \text{constant in the region of } 0.06).$$

After retardation, the energy of these electrons becomes

$$E_2 = fE_1 \qquad \left(f = (E_1 - eV_0)/E_1\right)$$

Where $V_0$ is the positive bias potential applied to the sample. The resolution therefore becomes

$$\Delta E_f = RE_2 \quad \text{or} \quad \Delta E_f = fRE_1.$$

Since $f < 1$, $\Delta E_f$ is smaller than $\Delta E_0$.

This device can be very effective. As well as improving the resolution of the spectrometer, it can extend the range of kinetic energy that the analyser can handle.

When using this device, the analyst must take account of a number of factors:

1) Only electrons emitted from the sample with an energy greater than $E_0$ will be detected. This means that some peaks will not be detected. It may be that, for some elements, a different Auger transition will have to be used.
2) The impact energy of the primary electron beam will be smaller by an amount $eV_0$. This will have some effect upon the emission cross section of the Auger electrons and it is likely to affect the spot size of the primary electron beam.
3) The arrangement shown in Figure 2.12 functions as an electrostatic lens which produces a virtual image of the sample below the sample. This changes the effective working distance and therefore the calibration and transmission of the analyser.
4) It is likely that the detected electrons will have left the sample at a larger emission angle than the standard CMA, changing the sampled depth.

None of these factors prevent the use of the device but the effect of all of them on the spectrum must be considered especially if quantitative information is being sought.

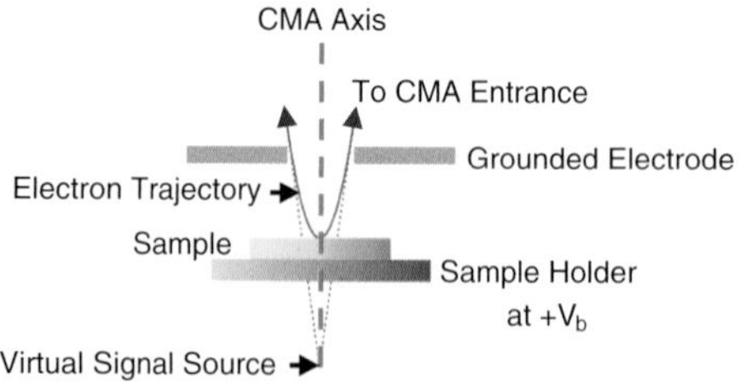


Figure 2.12 A method for improving upon the intrinsic energy resolution of the cylindrical mirror analyser (CMA).

### 2.5.2 The Hemispherical Sector Analyser

A HSA, also known as a concentric hemispherical analysers (CHA) or a spherical sector analyser (SSA), consists of a pair of concentric hemispherical electrodes between which there is a gap for the electrons to pass. Between the sample and the analyser there is usually a lens (transfer lens), or a series of lenses.

The lenses serve a number of purposes which will be discussed later but it is helpful to mention one now. The kinetic energy of the electrons as they are ejected from the sample is usually too great for the analyser to produce sufficiently high resolution, so they must be retarded. This retardation is achieved either within the lens or, using parallel grids, between the lens and the analyser. Retardation cannot occur in the field-free region between the sample and the lens entrance.

The schematic diagram of Figure 2.13 shows a typical HSA configuration for XPS. A potential difference is applied across the two hemispheres with the outer hemisphere being more negative than the inner one. Electrons injected tangentially at the input to the analyser will only reach the detector if their energy is given by

$$E = e\Delta V\left(\frac{R_1 R_2}{R_2^2 - R_1^2}\right)$$

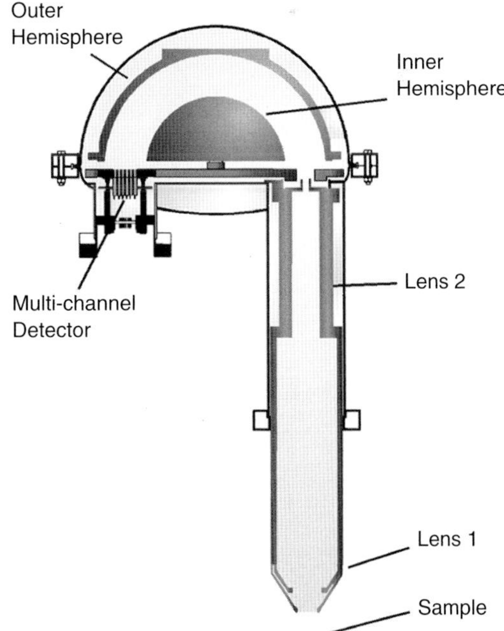


Figure 2.13 Schematic diagram of a modern hemispherical sector analyser (HSA) and transfer lens.

Where the kinetic energy of the electrons is given by $E$, e is the charge on the electron, $\Delta V$ is the potential difference between the hemispheres and $R_1$ and $R_2$ are the radii of the inner and outer hemispheres respectively. The radii of the hemispheres are constant and so the above equation can be expressed as

$$E = ke\Delta V$$

in which $k$ is known as the spectrometer constant and depends upon the design of the analyser.

An HSA also acts as a lens and so electrons entering the analyser on the mean radius will reach the exit slit even if they enter the analyser at some angle with respect to the tangent to the sphere given by the mean radius.

Electrons whose energy is higher than that given by the above expression will follow a path whose radius is larger than the mean radius of the analyser and those with a lower kinetic energy will follow a path with a smaller radius. Provided that the energy of these electrons does not differ too greatly from that given by the expression these electrons will also reach the output plane of the analyser. It is possible therefore to provide a number of detectors at the output plane. These detectors are arranged radially across that plane. Clearly, each of the detectors collects electrons of a different energy but by adding the signal into the appropriate energy channel, the sensitivity of the instrument can be increased by a factor equal to the number of detectors, if the size of the detector does not change. Instruments with up to nine discrete channel electron multipliers (channeltrons[3]) are commercially available. Some instruments are fitted with two-dimensional detectors which allow the user to select the number of channels collected at any one time. Up to 112 channels can be defined on some instruments, the width of each channel is small, so this does not imply that the sensitivity is 112 times that of an equivalent instrument fitted with a single channeltron. The advantage of having such a large number of channels is that it allows high-quality spectra to be recorded without scanning the analyser.

The HSA is typically operated in one of two modes; constant analyser energy (CAE) sometimes known as fixed analyser transmission (FAT) and constant retard ratio (CRR) also known as fixed retard ratio (FRR).

#### 2.5.2.1 CAE Mode of Operation

In the CAE mode electrons are accelerated or retarded to some user-defined energy which is the kinetic energy the electrons possess as they pass through the analyser (the pass energy). In order to achieve analysis in the CAE mode, the voltages on the hemispheres are scanned according to the graph shown in Figure 2.14.

3 'Channeltron' is a registered trademark of the Bendix Corporation. The word is in common usage meaning any channel electron multiplier.

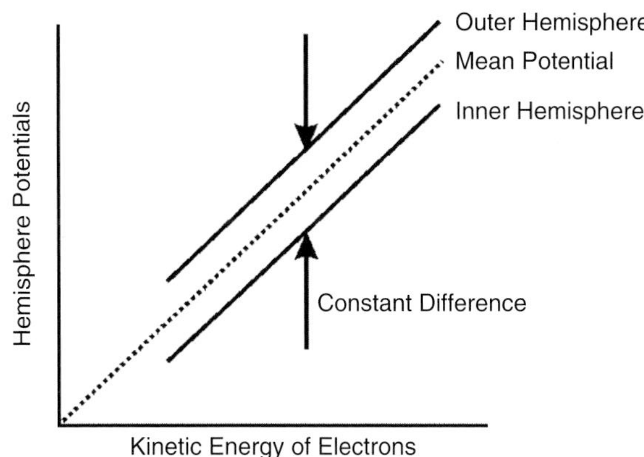


Figure 2.14 Operation of the hemispherical sector analyser (HSA) in constant analyser energy (CAE) mode.

The selected pass energy affects both the transmission of the analyser and its resolution. Selecting a small pass energy will result in high resolution whilst a large pass energy will provide higher transmission but poorer resolution. The pass energy remains constant throughout the energy range, thus, the resolution (in eV) is constant across the entire width of the spectrum.

The range of pass energies available to the user depends upon the design of the spectrometer but is typically from about 1 eV to several hundred electron volts. Figure 2.15 shows part of the XPS spectrum of silver (Ag 3d peaks)

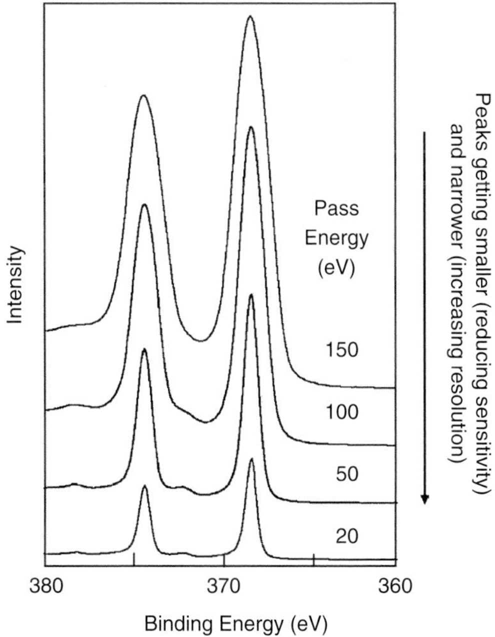


Figure 2.15 XPS of Ag showing the effect of pass energy upon the Ag 3d part of the spectrum.

recorded at a series of pass energies, showing the effect of pass energy on resolution and sensitivity. In a typical XPS experiment, the user will select a pass energy in the region of 100 eV for survey or wide scans and in the region of 20 eV for higher-resolution spectra of individual core levels. These high-resolution spectra are used to establish the chemical states of the elements present and for quantification purposes.

It is normal practice to collect XPS spectra in the CAE mode.

The energy resolution, ΔE, is dependent upon the pass energy ($E_p$), the width of the slit at the entrance to the analyser (W) and the angle at which the electrons enter the analyser in the dispersive direction (α);

$$\Delta E = E_p\left(\frac{W}{2R} + \frac{\alpha^2}{2}\right)$$

In some analyser designs, devices are included to limit the maximum value of α and thereby improve resolution at the expense of a small reduction in transmission.

#### 2.5.2.2 CRR Mode of Operation

In the CRR mode electrons are retarded to some user-defined fraction of their original kinetic energy as they pass through the analyser (the retard ratio). In order to achieve analysis in the CRR mode, the voltages on the hemispheres are scanned according to the graph shown in Figure 2.16.

In this mode, the pass energy is proportional to the kinetic energy:

$$\textit{Pass Energy} = \frac{\textit{Kinetic Energy}}{\textit{Retard Ratio}}$$

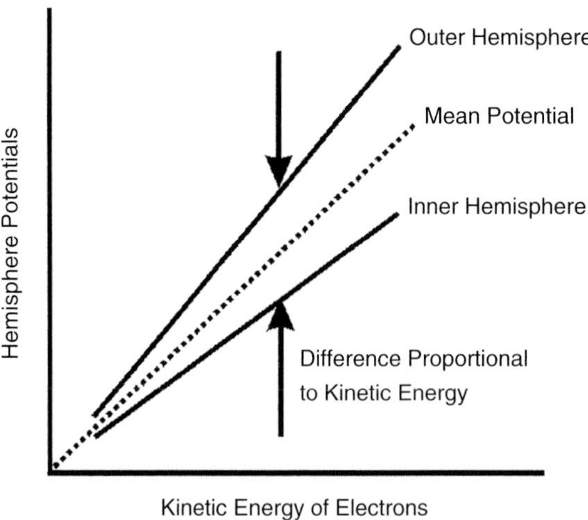


Figure 2.16 Operation of the hemispherical sector analyser (HSA) in constant retard ratio (CRR) mode.

The percentage resolution in this mode of operation is constant and inversely proportional to the retard ratio. The constant of proportionality depends upon the design of the instrument, on a typical instrument:

$$Resolution \approx (2/Retard\ ratio)\%$$

Typically, the resolution available from a good commercial HSA can be selected from the range 0.02–2.0%, in the case of the typical spectrometer referred to above, this means that the range of retard ratios is from 1 to 100.

It is normal practice to collect Auger electron spectra in the CRR mode, as this partially supresses the intense tail of low-energy secondary electrons. When collecting a survey spectrum or when only elemental information is required, a resolution in the region of 0.5% is often considered to provide a good compromise between sensitivity and resolution. If chemical state information is needed, then the resolution has to be better than this and a value in the range 0.02–0.1% is usually chosen depending upon the size of the chemical shift that needs to be observed.

Figure 2.17 shows the effect of changing the retard ratio on the Auger spectrum of aluminium metal which is covered with a thin layer of oxide.

If differential spectra are needed, these are usually calculated from the direct spectra within the data system, rather than using the phase sensitive detection approach that was popular in the past.

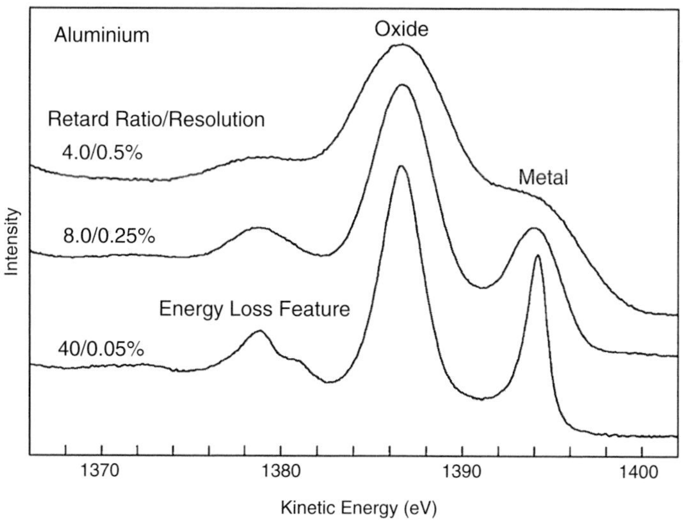


Figure 2.17 Auger KLL spectrum of aluminium showing the effect of changes in the retard ratio upon the resolution of the spectrum. At high retard ratio, the metal, oxide, and energy loss features become resolved.

#### 2.5.2.3 Comparison of CAE and CRR Modes

Figure 2.18 shows the XPS spectrum of copper collected in CRR and CAE modes. The retard ratio in Figure 2.18a is 4 and the pass energy in Figure 2.18(b) is 100 eV. This means that the transmission and resolution are identical at a kinetic energy of 400 eV (a binding energy of 1087 eV).

In the CRR mode, the pass energy is proportional to the kinetic energy and so the resolution, ΔE, becomes worse with increasing kinetic energy but the relative resolution, expressed as a percentage of the kinetic energy, ΔE/E, is constant throughout the energy range. The transmission increases with

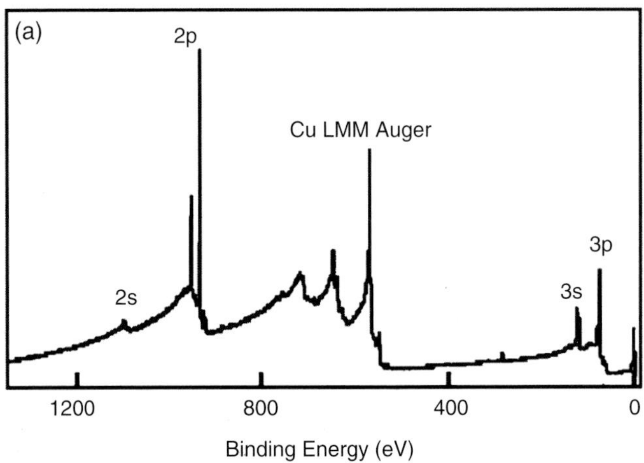


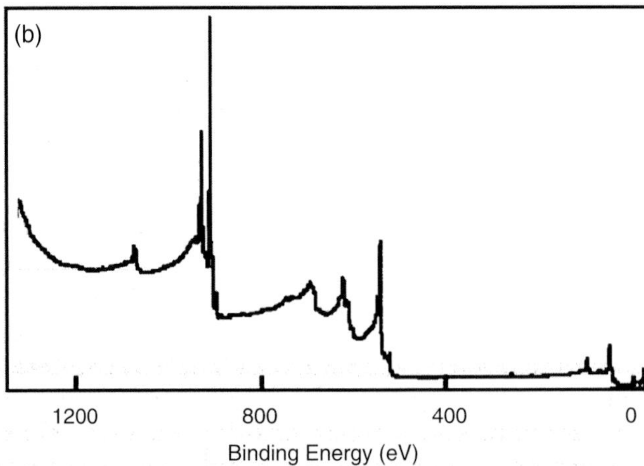


Figure 2.18 XPS survey spectra from copper acquired in (a) constant retard ratio (CRR) mode (retard ratio = 4) and (b) constant analyser energy (CAE) mode (pass energy = 100 eV).

increasing kinetic energy, which has the effect of suppressing the relatively high electron yield at the low kinetic energy end of the spectrum. This makes this mode of operation ideal for analysis using AES. As both spectral resolution and transmission change with electron energy, quantification of XPS spectra is difficult in the CRR mode of operation and CAE is preferred.

In the CAE mode of operation, the resolution, expressed in eV, and the transmission of the analyser both remain constant throughout the energy range of the scan. This ensures that XPS quantification is more reliable and accentuates the XPS peaks at the low kinetic energy (high binding energy) end of the spectrum.

#### 2.5.2.4 The Transfer Lens

The performance of the HSA is strongly dependent upon the nature and quality of the transfer lens or lenses between the sample and the entrance to the analyser. The presence of a transfer lens:

- Moves the analyser away from the analysis position allowing other components of the spectrometer to be placed closer to the sample.
- Maximises the collection angle to ensure high transmission and sensitivity.
- Retards the electrons prior to their injection into the analyser whilst allowing the sample to remain in an electrostatic field-free region.
- Determines and controls the area of the sample from which electrons are collected, allowing SAX measurements to be made.
- Controls the acceptance angle. This has obvious applications in defining the angular resolution for angle resolved XPS. It is also important for small area and imaging XPS because the angular acceptance will determine the spatial resolution as well as the transmission.

In the past, these transfer lenses were exclusively electrostatic, but some instruments are now fitted with an electromagnetic magnetic immersion lens. The current through the electromagnet is scanned as a function of the kinetic energy of the electrons being analysed. When using this type of lens, the sample is situated within the magnetic field. A major advantage of this type of lens is that it can collect electrons from a very wide range of emission angles; typically, a collection angle of 90° can be achieved compared with about 25°, for a purely electrostatic system. This significantly improves the sensitivity of the instrument. Magnetic lenses can provide better spatial resolution than electrostatic lenses of the same focal length because their aberration coefficients are lower.

Magnetic lenses cannot be used for AES because the magnetic field would deflect the primary electron beam and affect the spot size at the sample surface. The extent of these effects would depend upon the magnetic field strength, which must be varied with the kinetic energy being analysed.

### 2.5.3 Calibration of the Electron Spectrometer Energy Scale

In order to carry out meaningful analysis by XPS and AES, it is essential that the energy scale of the spectrometer, i.e. the electron energy analyser, is properly calibrated. In principle, this is a very simple procedure using pure metals that have been cleaned by in-situ ion etching to reduce C 1s/C *KLL* and O 1s/O *KLL* to a negligible level. The XPS binding energy scale is referenced to the Fermi level whilst the kinetic energy scale required for AES is, by convention, referenced to the vacuum level, which is 4.50 eV above the Fermi level. An ISO standard provides exact values for XPS binding energies[4] and AES kinetic energies[5] and the procedures to be used for calibration.

XPS calibration is achieved using gold and copper for unmonochromated sources with the addition of silver if a monochromated Al Kα source is employed. Multiple metals are necessary in order to establish not only the fidelity of the energy scale in a single position but also to ensure linearity across the entire binding energy range. The recommended values are listed in Table 2.3.

Once the necessary spectra have been recorded, and repeated seven times, and the necessary calculations made, the spectrometer is adjusted, if required, to align the peaks with the binding energies of Table 2.3 as specified in the ISO standard, the spectrometer is in a calibrated condition.

In the case of an AES spectrometer, the procedure to follow is the same although one must consider the type of analysis being carried out (elemental identification or high-resolution AES for chemical state identification) and the mode of operation (direct or differential AES). Values for elemental identification work, referenced to the vacuum level, for both direct and differential AES spectra are provided in Table 2.4. The Au $M_5N_{6,7}N_{6,7}$ transition is weak

Table 2.3 Binding energy reference values for calibration of XPS spectrometers.

| | | Binding energy/eV | | |
|---|---|---|---|---|
| **Element** | **Core level** | **Al Kα** | **Mg Kα** | **Monochromated Al Kα** |
| Gold | Au $4f_{7/2}$ | 83.95 | 83.95 | 83.96 |
| Silver | Ag $3d_{5/2}$ | not used | not used | 368.21 |
| Copper | Cu $L_3VV$ | 567.93 | 334.90 | not used |
| Copper | Cu $2p_{3/2}$ | 932.63 | 932.62 | 932.62 |

4 ISO 15472: 2010, Surface Chemical Analysis – XPS – Calibration of Energy Scale.
5 ISO 17974: 2002, Surface Chemical Analysis – AES – Calibration of Energy Scales for Elemental and Chemical State Analysis.

Table 2.4 Kinetic energy reference values for calibration of AES spectrometers for elemental identification referenced to the vacuum level.

| Element | Auger transition | Kinetic energy/eV | |
|---|---|---|---|
| | | Direct spectra | Differential spectra |
| Copper | Cu $M_{2,3}VV$ | 58 | 60 |
| Copper | Cu $L_3VV$ | 914 | 915 |
| Aluminium | Al $KL_{2,3}L_{2,3}$ | 1388 | 1390 |
| Gold | Au $M_5N_{6,7}N_{6,7}$ | 2011 | 2021 |

particularly at low-beam energies and the Al $KL_{2,3}L_{2,3}$ should be used instead. This peak is also used for spectrometers that do not scan beyond 2000 eV.

When an Auger electron spectrometer is used at high resolution, to establish chemical shifts associated with a particular transition, for instance, values derived from the XPS binding energy reference values are used, as described in the ISO standard.

## 2.6 Near Ambient Pressure XPS

Within a decade of the appearance of commercial XPS systems, with their attendant vacuum requirements, attention had been applied to the manner in which bridging the so-called pressure gap could be achieved. In the first decade this took two very different approaches; in Uppsala, Hans Siegbahn (Kai Siegbahn's son) developed a number of experimental routes to the XPS analysis of liquids, immersing a wire or a truncated cylinder in the liquid of interest, combined with differential pumping ensuring the liquid film remained intact and was amenable to XPS analysis in the usual way. In the UK, gas–solid reactions were enabled using a simple reaction cell, continuously dosed with gas, sited above a large bore pumping system. Both systems incorporated an electron transfer lens in their design between sample and electron analyser. This provided a degree of differential pumping, a necessity bearing in mind the complexity of in-situ sample processing during the analysis. In this way, the XPS spectra of a number of simple organic liquids and solid–gas reactions at high pressure (up to ca. 0.5 mbar) have been obtained. Notwithstanding such a promising start, interest from both researchers and manufacturers tailed off in the ensuing years and XPS became entrenched as an UHV technique. Advances were made slowly, but the design of gas-cells and differentially pumped analysis systems improved until manufacturers again considered the design and commercial viability of high-pressure XPS.

In addition to the possibilities mentioned above, there is also a significant group of samples that have the potential to benefit from a spectrometer that operates at higher than HV ambient pressure. The group includes those which are not completely vacuum compatible, for example, samples with large quantities of retained water or organic solvent (food products, wood, and other natural products, paper, some polymers, biological, and biomedical samples). In this respect one can view a higher pressure XPS system in a similar manner to a variable pressure scanning electron microscope (VPSEM), one which can accommodate samples that are less than vacuum compatible with ease. The samples highlighted above can all be analysed in a conventional XPS system but special approaches are required. These may simply rely on extended periods of time for pumping or chilling the sample towards liquid nitrogen temperatures. In recent years, commercial systems have been launched that are focused on NAPXPS systems (described variously as ambient pressure XPS or high-pressure XPS).

The cornerstone of commercial NAPXPS systems is the efficient design of a differential pumping arrangement between sample and the entrance slits of the hemispherical electron energy analyser. Such an arrangement will typically have up to four pumping stages (equipped with turbomolecular pumps) so the pressure is gradually reduced from the high pressure in the environs of the sample to the more usual UHV level at the analyser. This type of arrangement can be seen very clearly in the schematic of Figure 2.19, which represents the analyser configuration of one of the commercially available instruments.

Undoubtedly, the driving force to develop NAPXPS to its current level of sophistication and commercial viability has been that part of the surface science community involved with heterogeneous catalysis. The eventual goal being to analyse samples in their usual working environments, an achievement which has been termed *operando* XPS. Bearing in mind that many chemical processes take place at extremely high pressures, this goal is still some way off but it is clear that recent advances in NAPXPS have gone a considerable way towards bridging the pressure gap.

The benefits of using NAPXPS for samples with a high-vapour pressure (often referred to as 'gassy' samples) is very clear when one considers the processes that occur around the sample, illustrated in Figure 2.20.

Residual gas molecules around the sample are also ionised generating both electrons and positive ions. These are then able to impinge on the sample providing satisfactory charge neutralisation without the need for an additional flood gun. In this manner, charge compensation is achieved in NAPXPS in the same manner as VPSEM where insulating samples are often imaged without the usual expedient of applying a conducting coat.

An alternative approach, yet to be realised commercially, avoids the use of a complex (and expensive) differential pumping set-up. The sample, which could be solid or liquid is confined within a small high-pressure cell equipped with an

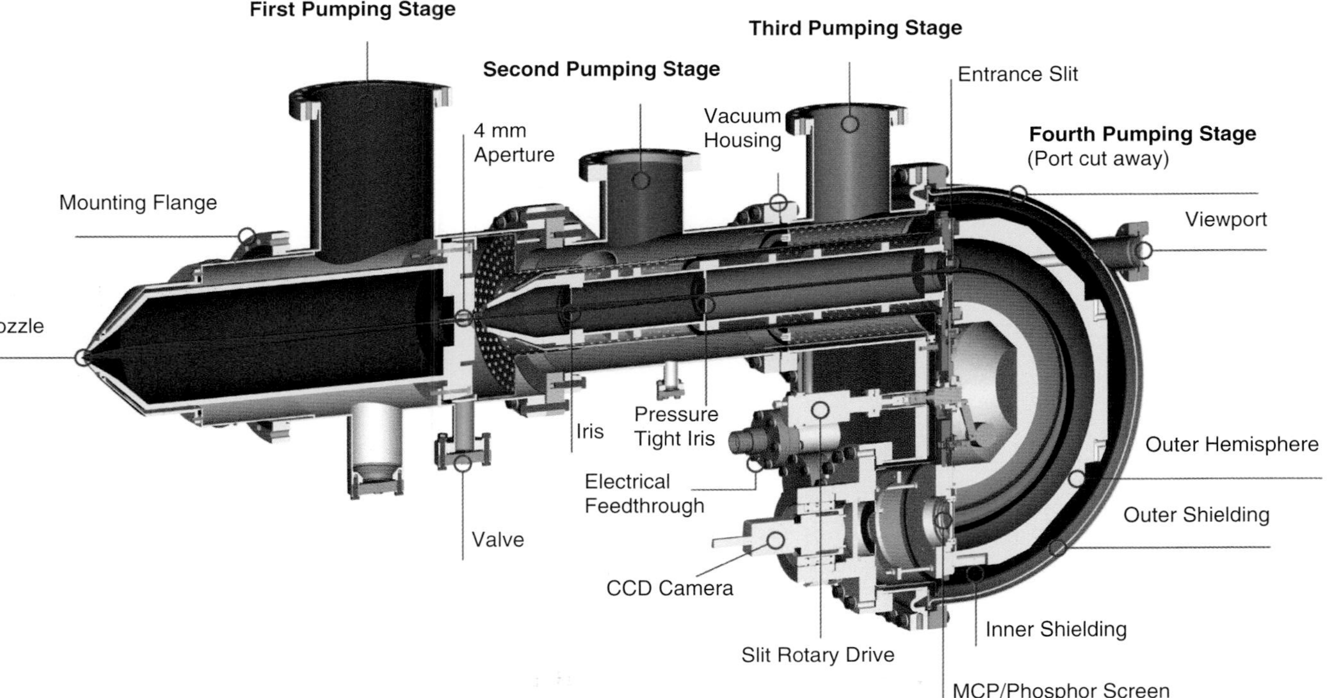


Figure 2.19 Schematic representation of the arrangement of the transfer lens and hemispherical analyser for near ambient pressure XPS (NAPXPS). The four differential pumping stages are clearly seen.

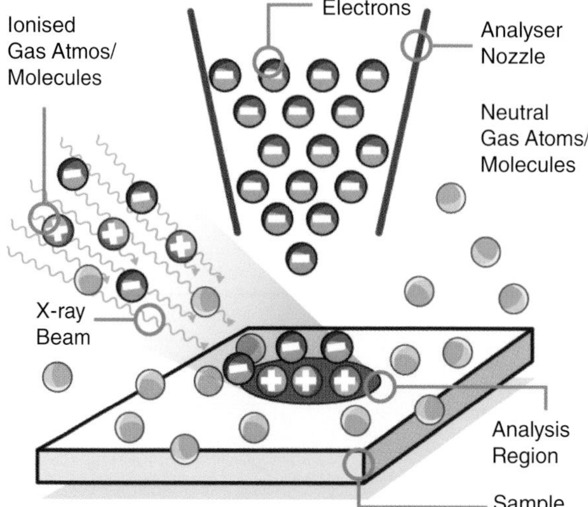


Figure 2.20 Charge compensation that occurs in near ambient pressure XPS (NAPXPS).

electron transparent, but molecularly opaque, window fabricated from graphene. The sample within the cell is irradiated with X-rays and a proportion of the photoelectrons that are generated pass through the graphene window and into the transfer lens of the spectrometer, which is maintained at the usual UHV conditions. In principle, such an arrangement could be installed in a conventional XPS instrument equipped with the appropriate connections within the analysis chamber for gas or liquid dosing of the cell. Such an arrangement is conceptually quite straightforward on a high-end XPS chamber, requiring simply an external gas or liquid handling arrangement and feed-throughs that could be arranged on a small Conflat flange.

## 2.7 Detectors

In most electron spectrometers, it is necessary to count the individual electrons arriving at the detector. To achieve this, electron multipliers are used. Although there are many types of electron multiplier, only two types are commonly used in electron spectrometers; channel electron multipliers (channeltrons) and channel plates.

### 2.7.1 Channel Electron Multipliers

These consist of a spiral-shaped glass tube with a conical collector at one end and a metal anode at the other. The internal walls of the detector are coated with a material which, when struck by an electron having more than some

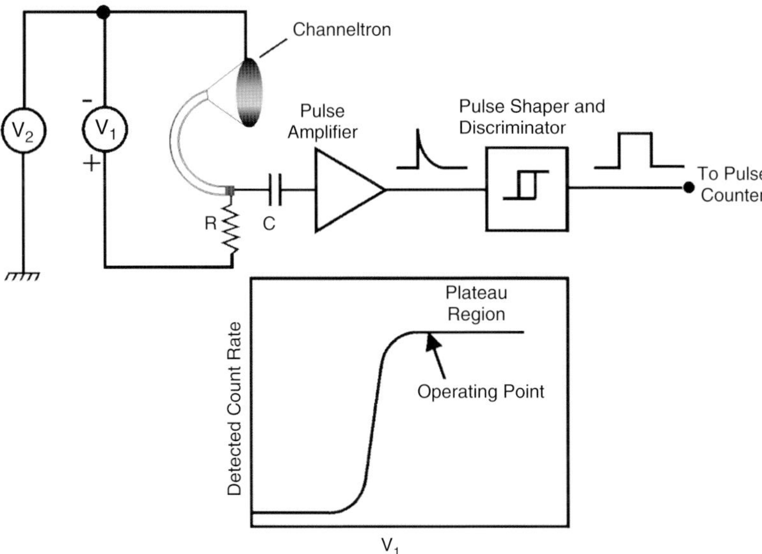


Figure 2.21 Schematic diagram showing the operation of a channel electron multiplier.

threshold kinetic energy, will emit many secondary electrons. A large potential difference is applied across the length of the channeltron, the cone being positive (Figure 2.21).

As an electron strikes the internal surface of the cone, electrons are emitted and accelerated into the tube of the detector where more collisions take place with the total number of electrons in the cascade increasing with each collision. The gain of a channeltron depends upon the potential difference between its ends. When the voltage is low, no output pulses are detected. As the voltage increases above some threshold, pulses can be detected. The detection efficiency continues to increase with voltage until a plateau voltage is reached. Above this voltage, the measured output count rate is independent of the voltage across the channeltron. The operating point should be chosen to be just above the plateau. The precise operating voltage will depend upon the age of the channeltron and its design. Typically, it will be between 2 and 4 kV. It is necessary to check periodically that the voltage across the channeltron is correct.

Each electron arriving at the detector typically results in about $10^8$ electrons reaching the anode. It is necessary to amplify these pulses of charge using a pulse amplifier and produce a square wave which can be counted by a rate metre. Use of a discriminator eliminates noise signals emanating from the channeltron or the preamplifier.

The channeltron samples about 5 mm in the dispersive direction of the analyser and about 15 mm in the non-dispersive direction. In order to increase the sensitivity of the spectrometer it is common to use an array of channeltrons as the detector. The channeltrons are arranged along the dispersive direction and so each one collects a different electron kinetic energy. The data system sums the output from each channeltron after applying the appropriate energy shift. Channeltrons can be capable of detecting up to about three million counts per second. They become non-linear at very high count rates but their range can be increased by calibrating them in the non-linear region. Above the non-linear region, they saturate and the output count rate falls to practically zero.

### 2.7.2 Microchannel Plates

A microchannel plate (MCP) is a disc with an array of small holes. Each of these holes behaves as a small channeltron. The gain of an individual channel is much lower than that of a channeltron so it is common to use a pair of MCPs in tandem. The maximum count rate which can be detected using an MCP is about 300 kcps on current systems for two-dimensional detection, but, in principle, could be as high as 10 Mcps for simple spectroscopy.

MCPs are used when it is necessary to detect data in two dimensions. Spectrometers have been designed using channel plates to measure signals:

- In an X-Y array for parallel acquisition of photoelectron images.
- In an X-Energy array for parallel acquisition of XPS line scans.
- In an Energy-Angle array for the parallel acquisition of angle-resolved XPS spectra.

### 2.7.3 Two-Dimensional Detectors

In a two-dimensional electron detector, not only is it necessary to detect the individual electrons leaving the analyser, but it is also necessary to detect the position that each electron strikes the detector. Figure 2.22 illustrates how this is usually accomplished on an XPS instrument.

Figure 2.22a shows the arrangement of the MCPs and the position-sensitive detector (PSD). An electron is shown striking the front face of one of the MCPs. Figure 2.22b is a magnified portion of the first MCP and illustrates how the input electron strikes the wall of one of the channels, causing many electrons to be emitted, each of which also strikes the channel wall causing further emission and resulting in a 'cloud' or pulse of electrons leaving the back of the MCP. These electrons then strike the front of the second MCP resulting in a much larger pulse of electrons leaving the back of the second MCP and striking the PSD.

There are two types of PSD in common use for XPS, the resistive-anode detector (RAD) and the delay-line detector (DLD).

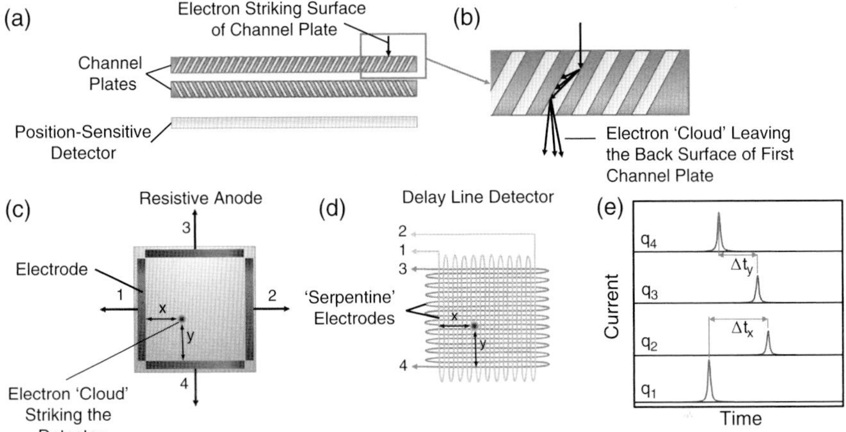


Figure 2.22 Schematic diagram illustrating the way in which channel plate detectors are used in conjunction with a position-sensitive detector (PSD); (a) a cross section of the detector showing the relative positions of a pair of channel plates and a PSD; (b) Magnified portion of the first channel plate; (c) A resistive-anode PSD; (d) A PSD based upon a pair of delay lines; (e) Illustration of the signal coming from the PSD output leads. A qualitatively similar type of output would be observed for each type of PSD.

#### 2.7.3.1 The Resistive-Anode Detector

The RAD illustrated in Figure 2.22c is a square device having an electrode along each side and a resistive material between them. Other geometries are possible. When a pulse of electrons arrives at a point on the detector, charge flows to each of the electrodes. The charge is measured at each electrode and the location of the incident pulse of electrons can be computed:

$$x = k_x \frac{q_1 - q_2}{q_1 + q_2} \quad \text{and} \quad y = k_y \frac{q_s - q_4}{q_s + q_4}$$

Where $k_x$ and $k_y$ are scaling factors based on the geometry of the device and $q_n$ is the charge that arrives at electrode n. The detector output signals are illustrated in Figure 2.22e. The magnitude of the signals resulting from each pulse is measured but, in this type of detector, the arrival time of the pulse at each electrode is not taken into account.

#### 2.7.3.2 The Delay-Line Detector

The Delay-Line Detector (DLD) is illustrated in Figure 2.22d. It consists of two lengths of fine wire, each wound into a 'serpentine' path. The two 'serpentines' are superimposed orthogonally so that one can be used to measure the position of a pulse of electrons in the x direction whilst the other measures in the y direction. Again, when a pulse of electrons arrives at a point on the detector, charge flows to each of the four electrodes (1, 2, 3, and 4) and the pulse is detected at each electrode. It is the relative time of arrival of the pulses ($\Delta t_x$ and $\Delta t_y$ in Figure 2.22e) that provides the positional information. In this case, the magnitude of the pulses at each electrode is not taken into account.

## 2.8 Small Area XPS

It is often necessary to analyse a small feature or imperfection on the surface of a sample. For the analysis to be effective, as much as possible of the signal from the surrounding area should be excluded. This is usually done in one of two ways:

1) By flooding the analysis area with X-rays but limiting the area from which the photoelectrons are collected using the transfer lens. This is described as lens-defined small area analysis.
2) By focusing a monochromated beam of X-rays into a small spot on the sample, source-defined small area analysis.

### 2.8.1 Lens-Defined Small Area XPS

In most spectrometers, the transfer lens fitted to the analyser is operated in such a way as to produce a photoelectron image at some point in the electron optical column. If a small aperture is placed at this point, then only electrons emitted from a defined area of the sample can pass through the aperture and reach the analyser. If the magnification of the lens is $M$ and the diameter of the aperture is $d$, then the diameter of the analysed area is $d/M$. In some instruments, an aperture can be selected from a number of fixed apertures whilst in other instruments an iris is used to provide a continuous range of analysis areas.

Spherical aberrations, which occur in any electron optical lens system, mean that the acceptance angle of the lens has to be limited to provide good edge resolution in the analysis area. This is achieved using either another set of fixed apertures or another iris placed at some point remote from the image plane of the lens.

Using this technique, commercial instruments can provide small area analysis down to <20 μm.

This is an effective method for producing high-quality, small-area XPS data but it suffers from a disadvantage. Reducing the angular acceptance of the lens reduces the detected flux per unit area of the sample. This means that analysis times can become very long. During the analysis time the whole of the sample (or samples in a multi-sample experiment) is being exposed to X-rays, potentially resulting in radiation damage. Therefore, if many samples or many points on a single sample are to be analysed then the analyst cannot be certain that the surface remains unaltered.

### 2.8.2 Source-defined Small Area Analysis

It is possible to shape a quartz crystal so that it can focus a beam of X-rays *and* provide monochromatic X-rays by diffraction. In this respect, it behaves rather like a concave mirror. The focusing is usually achieved using a magnification of unity which means that the size of the X-ray spot on the sample is approximately equal to the size of the electron spot on the X-ray anode. Analysis areas down to about 10 μm can be achieved in commercial instruments using this method.

Because the source of X-rays is defining the analysis area, aberrations in the transfer lens will not affect the analysis area and so the lens can be operated at its maximum transmission, regardless of how small the analysis area becomes. The sensitivity of a spectrometer operating in this mode is therefore much higher than that of an equivalent instrument operating in the lens defined mode. This reduces the danger of sample damage during analysis and eliminates radiation damage to the surrounding area of the sample(s).

The intensity of the X-ray beam is proportional to the intensity of the electron beam at the anode. The intensity of the electron beam at the anode is limited by the rate at which heat can be removed from the vicinity of the electron spot. The X-ray source can be operated at several hundred watts for large area analysis whilst it is only possible to use a few watts at the smallest areas.

## 2.9 XPS Imaging and Mapping

A logical extension to SAX is to produce an image or map of the surface. Such an image or map shows the distribution of an element or a chemical state on the surface of the sample. There are two distinct approaches, used by manufacturers, to obtaining XPS maps. These are:

1) Serial acquisition in which each pixel of the image is collected in turn.
2) Parallel acquisition in which data from the whole of the analysis area is collected simultaneously.

### 2.9.1 Serial Acquisition

Serial acquisition of images is based on a two-dimensional, rectangular array of SAX analyses. By this means, the distribution of elements or chemical states can be measured. The ultimate spatial resolution in the image is determined by the size of the smallest analysis area (this depends upon the instrument but 10 μm is possible using a source-defined approach with a high-quality, modern spectrometer). Serial acquisition is generally slower than parallel acquisition but has the advantage that one can collect a range of energies at each pixel whereas, in parallel acquisition, only a single energy can be collected during each acquisition.

There are several methods by which the analysis area can be stepped over the field of view of the image:

1) *Scanning the sample stage.* Using this method, the analysis position is fixed in space and the sample surface is moved with respect to this position. The advantages of this method are that all of the important instrumental conditions remain constant (e.g. the energy of the X-ray beam, the resolution of the analysis spot and the transmission function of the lens) and the maximum size of the image field of view is limited only by the range of motion of the sample stage. The disadvantages of the method are that it tends to be slower than other methods and requires a high-precision stage with low backlash.

   Traditionally, this type of mapping involves moving the stage to the required position and allowing a settling time before collecting the data and moving the sample to the next position. The settling time can be eliminated by using a method of continuous scanning of the sample stage during the acquisition. There is at least one commercially available instrument that provides this capability.
2) *Scanning the lens.* This method requires that two pairs of deflector plates be built into the lens. By applying potentials to these plates, the photoelectron image can be deflected with respect to the area-defining aperture within the transfer lens. The analysis area can therefore be scanned in the X and Y directions and a map built up.

   The advantage of this method is that it is faster than scanning the stage but it suffers from a major disadvantage. The resolution of the map rapidly degrades as a function of distance from the centre of the map due to the spherical aberrations which are inevitably present in the electrostatic lens.
3) *Scanning the monochromated X-ray spot.* As mentioned in the discussion of SAX, it is possible to produce an X-ray image of the spot of electrons on the X-ray anode. Therefore, if the spot of electrons is scanned on the anode, the X-ray beam will be scanned on the sample surface. Again, the transfer lens

can be operated in its maximum transmission mode because it does not contribute to the spatial resolution.

The advantages of this method are the same as those for source-defined SAX. The disadvantage is that the field of view is very limited. This is especially true in the direction of the Bragg angle. As the electron beam is scanned over the anode, the diffraction angle is changing and so the wavelength of the X-rays reaching the sample is changing. In turn, this means that the kinetic energy of the photoelectrons will change and so the energy to which the analyser is tuned must be adjusted as a function of distance. If the deflection angle is too large there will be no intensity in the Al K$\alpha$ radiation at the required wavelength and so photoemission is not possible. If it is necessary to map large areas of the sample, this method must be used in combination with stage scanning.

### 2.9.2 Parallel Acquisition

In parallel acquisition of photoelectron images, the whole of the field of view is imaged simultaneously without scanning voltages applied to any component of the spectrometer.

There are two types of spectrometer, available commercially, which are capable of parallel imaging as well as conventional XPS spectroscopy. One has an additional lens and detector at the exit from a normal hemispherical analyser. The other has a spherical mirror analyser (SMA), used for imaging, in tandem with the hemispherical analyser which is used for spectroscopy. In the latter case, a single detector is used for both imaging and spectroscopy.

#### 2.9.2.1 Parallel Imaging Using a Hemispherical Spectrometer

Figure 2.23 shows schematically how this type of spectrometer works. After leaving the sample, the photoelectrons pass through lenses 1 and 2 in the transfer lens assembly, producing a photoelectron image of the sample surface at some plane within the lens column after each lens (the image planes). Up to this point, the lens system is very similar to a conventional spectrometer. If the spectrometer is to be capable of parallel imaging, a third lens (Lens 3) must be placed between Lens 2 and the analyser. Lens 3 is operated such that its focal length is equal to the distance between the lens and the second image plane. This means that electrons emanating from any one point on the image will leave lens 3 on parallel paths. The angle between the beam of electrons and the lens axis will depend only upon their distance from the lens axis in the image plane. The electrons then enter the analyser, which functions as both an energy filter *and* a lens. The deflection angle of the analyser is 180° and so the angular distribution of the electrons at the input to the analyser is retained at the exit. This means that a fourth lens (Lens 4), operated in the reverse manner to that

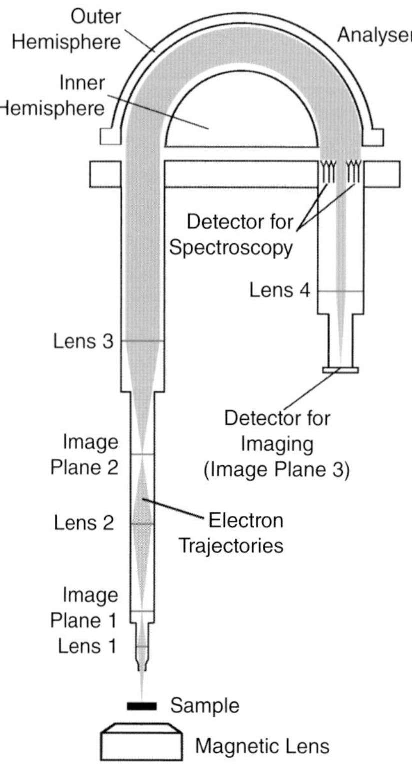


Figure 2.23 Schematic diagram illustrating the principles of parallel acquisition of photoelectron images.

of Lens 3 will reconstruct the photoelectron image at a two-dimensional detector placed at the focal length of lens 4.

#### 2.9.2.2 Parallel Imaging Using a Spherical Mirror Analyser

Figure 2.24 shows the analyser arrangement for this type of spectrometer. At the heart of the analyser is a conventional hemispherical analyser. This is surrounded by a SMA. At the input to the analyser is a set of lenses analogous to Lens 1 and Lens 2. A magnetic immersion lens is often used with both types of analyser. For spectroscopy, the HSA is operated in the normal manner and the photoelectrons are detected on a two-dimensional detector placed at the exit from the analyser. For parallel imaging, there is no field applied between the inner and outer hemispheres and so the electrons travel in a straight line to a hole in the outer hemisphere. A field is present between the outer hemisphere and the SMA, and this causes the photoelectrons to be reflected and focused on the two-dimensional detector after passing through another hole in the outer hemisphere. A baffle, placed in the SMA region of the analyser provides an energy filter.

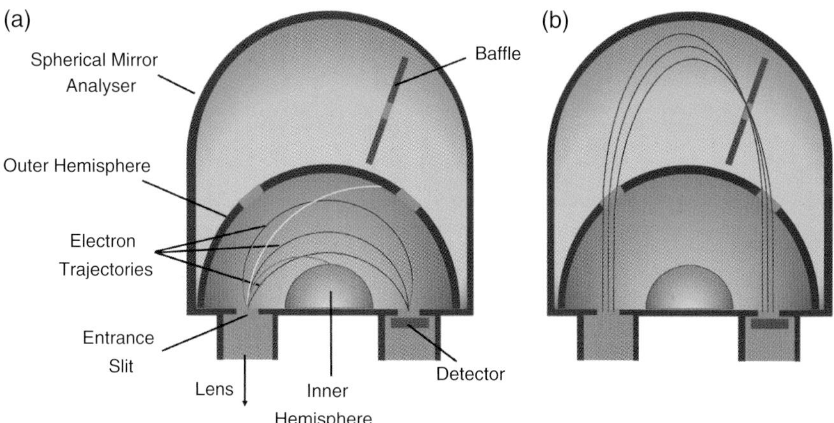


Figure 2.24 Parallel imaging using a spherical mirror analyser (SMA). (a) Electron trajectories when the analyser is used for spectroscopy; (b) Electron trajectories when the analyser is used in the parallel imaging mode. The trajectories shown in red are those of electrons having a kinetic energy equal to the pass energy of the analyser, yellow indicates a kinetic energy greater than the pass energy and blue a kinetic energy less than the pass energy.

#### 2.9.2.3 Spatial Resolution and Chemical Imaging

The spatial resolution of parallel imaging is dependent upon the spherical aberrations in the lens. Limiting the angular acceptance of the lens can reduce the effect of the aberrations and so resolution can be improved at the expense of sensitivity. The use of a magnetic immersion lens in the sample region also reduces aberrations and therefore allows higher sensitivity at a given resolution. This method of imaging is relatively fast and commercial instruments can produce images with an image resolution of $< 3\,\mu m$.

Parallel imaging clearly provides the best image resolution, but it only collects an image at a single energy. It is customary to make a measurement at a photoelectron peak energy and a second measurement at some energy remote from the peak where the signal intensity is approximately equal to the estimated background signal under the peak maximum. By subtracting the background signal from the signal at the peak maximum, a more accurate measurement can be made. This is in contrast with the use of a serial mapping method in conjunction with a multi-channel detector, to produce a 'snapshot' spectrum at each pixel of the map. Such spectra can then be treated with advanced data processing techniques to extract the maximum chemical information from the image.

To acquire a complete spectral array with spectrometer operating in a parallel imaging mode, it is necessary to collect a series of images with small binding energy increments and post process the desired pixel array to achieve a high-resolution XPS spectrum. Figure 2.25 shows two examples of this type of

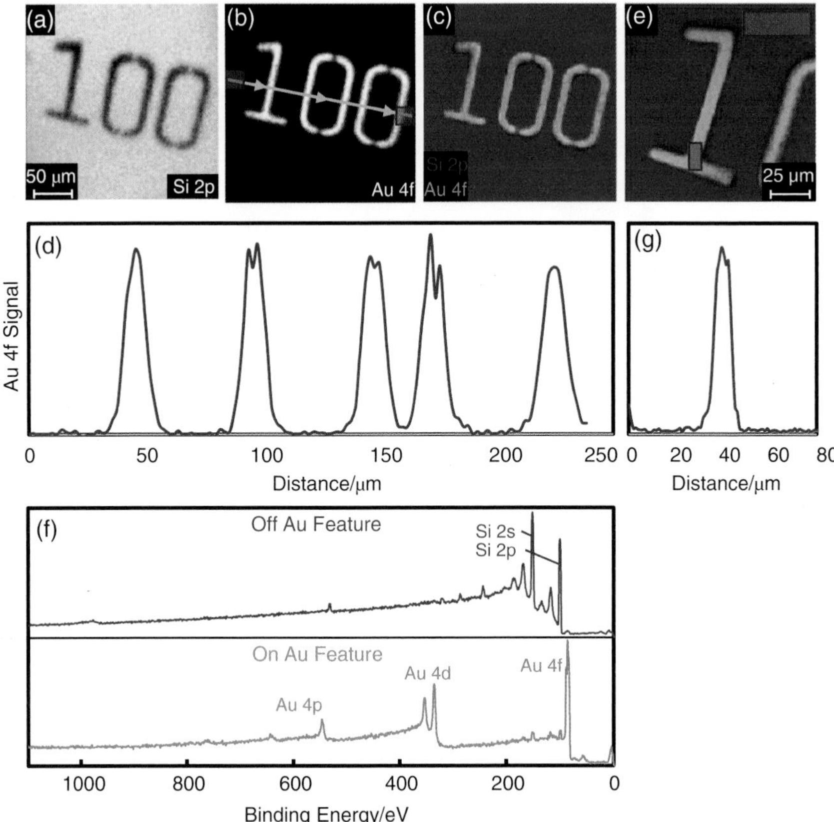


Figure 2.25 Examples of parallel XPS images from gold features on a glass substrate. (a) An image from a 250 µm× 250 µm area of the sample using electrons whose binding energy was 103 eV (Si 2p from $SiO_2$). (b) Similar to (a) but using electrons having a binding energy of 84 eV (Au 4f). (c) an overlay of the images in (a) and (b). (d) A line scan through the Au 4f image, along the orange line shown in (b). (e) as (c) but at a higher magnification. (g) An Au line scan across the '1' feature shown in (e). (f) Spectra, derived from the image stack, from the areas outlined in blue in (e).

measurement where the images were collected from a sample consisting of a glass substrate with gold features. A parallel image was collected at 1 eV intervals from 0 to 1104 eV binding energy. Figure 2.25a shows the image collected at a binding energy of 103 eV (the binding energy of Si 2p in $SiO_2$). Figure 2.25b was collected at 84 eV (the binding energy of Au 4f) and Figure 2.25c is an overlay of these two images showing Si in red and Au in green. The total signal from the pixels contained within the blue square at the left side of Figure 2.25b was measured and the square then translated across the image in the direction shown by the orange line to produce a line scan, shown in Figure 2.25d.

The experiment was repeated at a higher image magnification to obtain Figure 2.25e and Figure 2.25g. Measurements on the feature shown in the line scan, Figure 2.25g, show that the edge resolution is better than 2 μm. Small area spectra were constructed from the pixels within the blue rectangles shown on Figure 2.25e, these spectra are shown in Figure 2.25f.

This is a method for obtaining spectra from very small areas. It is possible to obtain such spectra from irregular shapes over a feature of interest thus maximising the signal available.

To estimate the lateral resolution in SAX, the following practice may be adopted. Assume, initially, that the analysis area in a small area mode of acquisition is circular and, outside this area, there is no transmission. Transmission is uniform within the area, (i.e. the transmission as a function of position on the sample resembles a 'top hat' distribution). To determine the dimensions of the analysis area a knife-edge sample (often silver) is translated through the analysis area whilst measuring the XPS signal from the knife-edge (e.g. the signal from the Ag $3d_{5/3}$ peak), see Figure 2.26. The signal is initially zero until the knife-edge begins to intercept the analysis area when it begins to rise. The

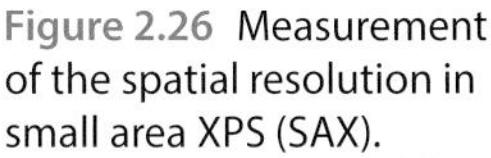

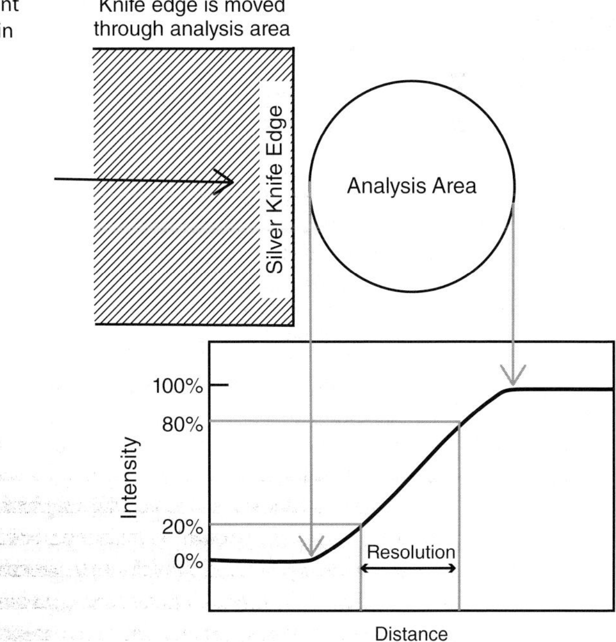


Figure 2.26 Measurement of the spatial resolution in small area XPS (SAX).

intensity of the signal continues to rise until the silver completely fills the whole of the analysis area.

The signal is then plotted against distance, as shown in Figure 2.26. The distance through which the knife-edge must be translated for the signal to change between two prescribed percentages of the total signal change is then determined. This distance is then described as the lateral resolution for spectroscopy. The percentages used depend upon the instrument manufacturer. The range 20–80% is used by some because the reported spatial resolution is then approximately equal to the radius of the analysis area (within < 2%). If the response of the spectrometer can be accurately represented by a Gaussian curve, instead of a top hat, then a value of 16–84% represents the distance between the two points, each one standard deviation on either side of the centre of the analysis area.

The method described above is specifically for the case when the analysis area is defined by the transfer lens ahead of the analyser (lens-defined SAX). The same method can be used when the area is defined by the X-ray spot size (source-defined SAX). Measurement of spatial resolution should be made in two orthogonal directions in case the analysis area is not circular. Refer to ISO 18156 for full experimental details on this method.

## 2.10 Angle Resolved XPS

As mentioned in the previous chapter, the finite mean free path of electrons within a solid means that the information depth in XPS analysis is of the order of a few nanometers. This, of course, is only true if the electrons are detected at a direction normal to the sample surface. If electrons are detected at some angle to the normal, the information depth is reduced by an amount equal to the cosine of the angle between the surface normal and the analysis direction. This is the basis for a powerful analysis technique, angle-resolved XPS (ARXPS). One of the reasons for the usefulness of the method is that it can be applied to films which are too thin to be analysed by conventional depth profiling techniques or those that would be irretrievably damaged by such methods. Another reason to use ARXPS is that it is a non-destructive technique which can provide chemical state information, unlike methods based upon sputtering using monatomic ions.

To obtain ARXPS data, the angular acceptance of the transfer lens is set to provide good angular resolution, usually the half angle is set to be in the region of 1°–3°. A series of spectra is then acquired as the sample surface is tilted with respect to the lens axis. Figure 2.27 illustrates how the spectra might appear at each end of the angular range if the sample consists of a thin oxide layer on a metal substrate. Note that the relative intensity of the oxide peak is larger at the near grazing emission angle.

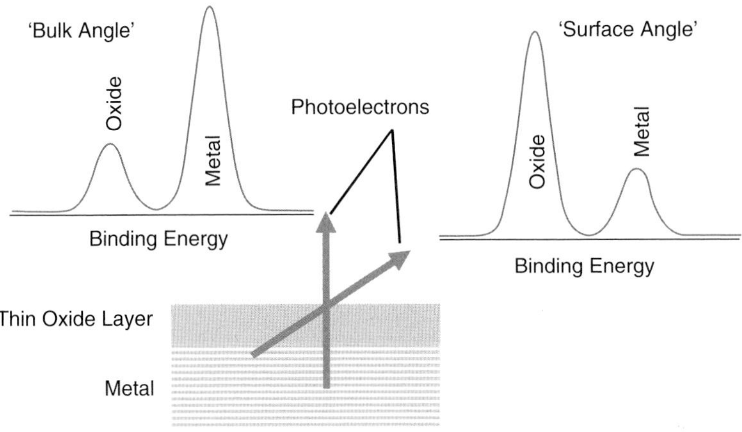


Figure 2.27 Illustration of XPS spectra taken from a thin oxide film on a metal at near normal collection angle and near grazing collection angle.

There are commercially available instruments that produce ARXPS spectra without tilting the sample; they are capable of parallel collection of angle-resolved data. This has several advantages over the conventional method:

- ARXPS measurements can be taken from very large samples, such as complete semiconductor wafers. Such samples are too large to be tilted inside an XPS spectrometer.
- The analysis position remains constant throughout the angular range. When combining SAX and ARXPS it is difficult to ensure that the analysis point remains fixed during the experiment, especially if that point is remote from the tilt axis even with a eucentric stage. Since the sample does not move during parallel collection, the analysis position remains constant.
- The analysis area remains constant during the analysis. If lens-defined SAX is combined with ARXPS then the analysis area would increase by a large factor as the sample is tilted away from its normal position. Using a combination of source-defined small area analysis and parallel collection, the analysis area becomes independent of angle.

The use of ARXPS as a non-destructive method for near-surface depth profiling of samples will be discussed further in Chapter 4.

Figure 2.28 shows the arrangement within an instrument capable of parallel collection of angle-resolved photoelectrons. The two-dimensional detector at the output plane has photoelectron energy dispersed in one direction (as with a conventional lens analyser arrangement) and the angular distribution dispersed in the other direction. Such an arrangement can provide an angular range of 60° with a resolution close to 1°.

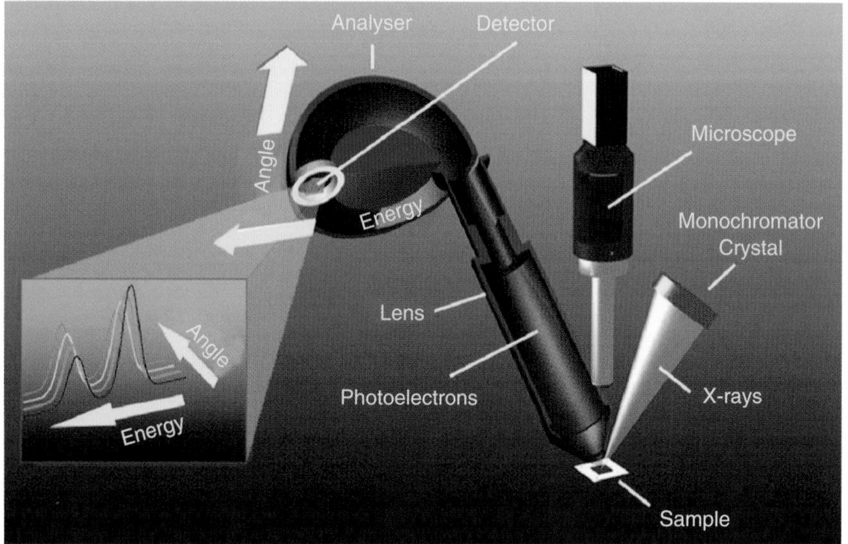


Figure 2.28 The arrangement of a spectrometer capable of collecting angle resolved spectra in parallel.

## 2.11 Automation

XPS has had the reputation for being an analytical technique that required expert operators. An analytical laboratory would have only a small number of operators (frequently, only one) capable of using the spectrometer. The complexity of the instrumentation was preventing these laboratories from training scientists to analyse their own samples. This, coupled with the cost of the XPS equipment, was a reason for some laboratories to use specialist, third-party analysis companies for their XPS needs.

This is now changing. There are instruments that can be operated by users after only a very small amount of training, lasting hours rather than days. The operator is required to mount the sample(s) on a sample holder, place the holder in the load lock and choose a 'recipe' for the analysis. The computer-controlled instrument then pumps the load lock to the required pressure and transfers the sample to the analysis chamber. At this point, the operator would have to choose the analysis position(s) and allow the chosen recipe to run. Items in an XPS recipe could include spectroscopy, line scans, maps, depth profiles and ARXPS measurements, with the computer controlling all the hardware and selecting the appropriate conditions for the required analysis. Once the measurements have been made, a report, which includes spectra, peak identification and quantification tables, may be generated as part of the recipe. With an instrument of this type, analytical laboratories can train a

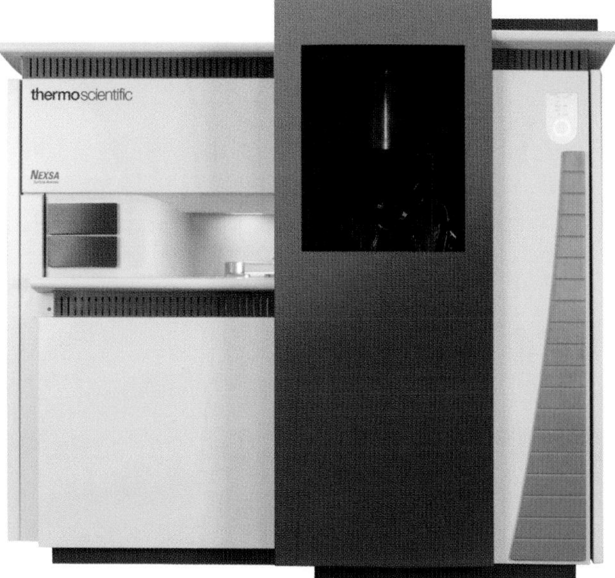


Figure 2.29 An example of a fully computer-controlled XPS instrument.

greater number of scientists to analyse their own samples. For other types of spectroscopy this has, of course, been possible for decades. A fully computer-controlled XPS instrument is shown in Figure 2.29.

Full automation allows the possibility of remote operation of the instrument over a computer network or via the internet, which is routinely carried out in The Surface Analysis Laboratory at the University of Surrey.

# 3

# The Electron Spectrum: Qualitative and Quantitative Interpretation

## 3.1 Introduction

The output from an electron spectrometer is amenable to many levels of interpretation. It ranges from a simple qualitative assessment of the elements present to a full-blown quantitative analysis complete with assignments of chemical states, and determination of the phase distribution for each element. In practice, a happy medium is usually required with an estimation made of the relative amounts of each element present. There are certain similarities in the way that AES and XPS spectra are treated. We shall initially consider them together as this also provides a means of comparing the analytical capabilities of the two methods.

## 3.2 Qualitative Analysis

The first step to be taken in characterising the surface chemistry of the sample under investigation is the identification of the elements present. To achieve this, it is usual to record a survey, or wide-scan spectrum over a region that will provide fairly strong peaks for all elements in the periodic table. For both XPS and AES, a range of 0–1000 eV is often sufficient. The current IUVSTA[1] recommendations for the acquisition of XPS survey spectra extend this range, as shown in Table 3.1. This is often known as the "St Malo Protocol" as recommendations were developed during the IUVSTA-sponsored workshop 'XPS from Spectra to Results,' held in St Malo in 2002.[2]

1 IUVSTA is the International Union for Vacuum Science, Technique, and Applications.
2 Castle, J.E. and Powell, C.J. (2004). *Surf. Inter. Anal.* 36: 225–237.

*An Introduction to Surface Analysis by XPS and AES*, Second Edition.
John F. Watts and John Wolstenholme.

Table 3.1 IUVSTA recommended conditions for the acquisition of an XPS survey spectrum.

| | Mg Kα | Al Kα |
|---|---|---|
| Binding energy range (eV) | 0–1150 | 0–1350 |
| Binding energy step size (eV) | 0.4 | 0.4 |

The spectrum should be recorded in the CAE mode of analyser operation at conditions that provide a resolution (FWHM) of 2.0 eV on the $Ag3d_{5/2}$ peak.

The individual peaks may be identified with the aid of data in tabular or graphical form as reproduced in Appendices 1 and 2 or using the peak identification routine included within the software of most commercial datasystems. Auger spectra may be recorded in either the direct (Figure 3.1a), or the differential (Figure 3.1b) mode. Nowadays, the direct mode is rather more popular particularly with high spatial resolution scanning Auger microscopes.

The photoelectron spectrum from a similar sample, Figure 3.1c is composed of the individual photoelectron peaks and the associated Auger lines resulting from the de-excitation process following photoemission. Unlike the Auger spectrum of Figure 3.1b the electron background is relatively small and increases in a step-like manner after each spectral feature. This is a result of the scattering of the characteristic Auger or photoelectrons within the matrix, bringing about a loss of kinetic energy. The shape of this background itself contains valuable information and, to the experienced electron spectroscopist, provides a means of assessing the way in which near-surface layers are arranged. In the case of a perfectly clean surface, the photoelectron peaks will have a horizontal background or one with a slightly negative slope. If the surface is covered with a thin overlayer, the peaks from the buried phase will have a positive slope. In the most severe case, the peak itself will be absent and the only indication will be a change in background slope at the appropriate energy.

The rising background at the low kinetic energy (high-binding energy) end of Figure 3.1d, acquired using a non-monochromatic Al Kα source, should be noted and compared with the background in Figure 3.1c. The former is a result of low-energy electrons generated by the bremsstrahlung component of the source, as it comprises of a broad range of photon energies it generates a similarly disperse range of electron energies. The other result of the bremsstrahlung component is the significantly more intense (by a factor of about two) O *KLL* Auger transition. There may also be a minor component of electrons emanating from the non-monochromatic X-ray source, although this is kept to a minimum by the use of a thin (< 5 μm) aluminium window between source and sample.

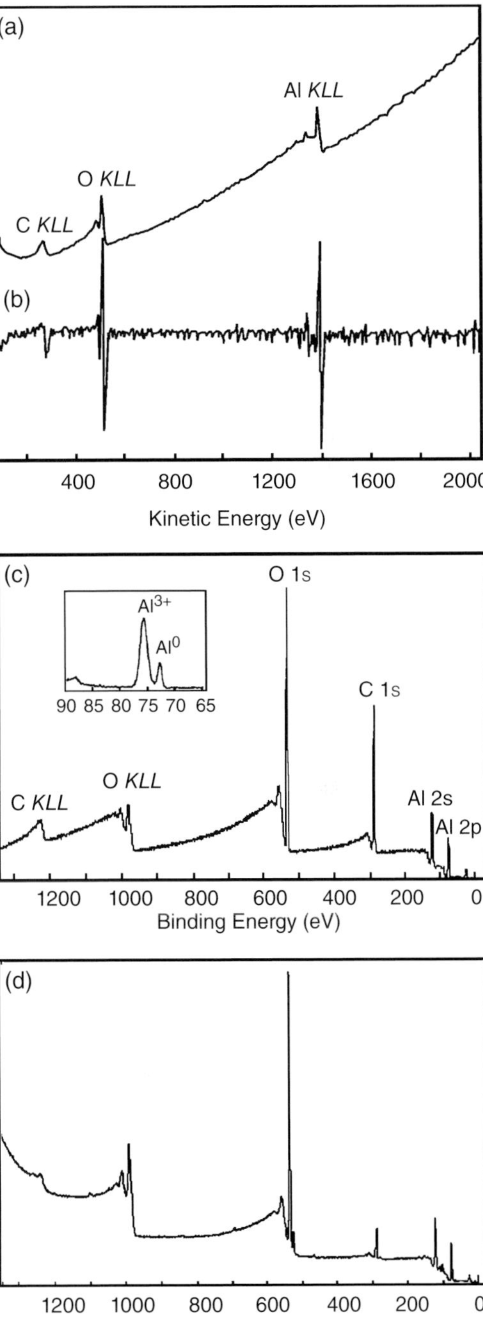


Figure 3.1 Electron spectra from air-oxidised aluminium foil. (a) a direct Auger spectrum; (b) a differential Auger spectrum; (c) an XPS spectrum recorded using a monochromated X-ray source; (d) an XPS spectrum recorded using a non-monochromated X-ray source. The inset in (c) is the Al 2p region recorded at higher resolution showing the metallic ($Al^0$) and oxide ($Al^{3+}$) components. Note the presence of carbon in all of these spectra; the result of the deposition of adventitious carbon from the atmosphere.

### 3.2.1 Unwanted Features in Electron Spectra

The XPS spectrum is further complicated by the presence of several features of no analytical use such as X-ray satellites and X-ray ghosts. X-ray satellites are present if non-monochromated radiation is used and occur because the characteristic transitions are excited by a minor component of the X-ray spectrum, e.g. Al K$\alpha_{3,4}$, Al K$\alpha_{5,6}$, Al K$\beta$. Such features are *to be expected* and only present difficulties if they fall at the same binding energy as an element present in very small concentrations. The solution may then be to change to another radiation (Mg K$\alpha$) as the separations are slightly different. The Al K$\alpha_{3,4}$, X-ray satellite is easily identified in the XPS spectrum of Figure 3.1d for the most intense photoelectron peak (O 1s), as a small peak at a binding energy of approximately 520 eV. Other examples of X-ray satellites may be seen in Figure 2.6 and, later, in Figure 3.8. Auger transitions present in an XPS spectrum do not show such satellite features and this provides a rapid means of distinguishing between the two as seen on the O $KL_{2,3}L_{2,3}$ peak. X-ray ghosts arise from unsuspected X-rays irradiating the sample. These may result from 'crosstalk' in a twin anode gun. Crosstalk is the generation of a small amount of characteristic X-radiation from the anode that has not been chosen for the analysis, due to a misalignment of the source. Alternatively, X-ray ghosts may arise from the exposed base material of a damaged anode (e.g. Cu L$\alpha$, hν = 929.7 eV). In either case, the problem should be reduced to an inconsequential level by overhauling and readjusting the X-ray gun.

The use of a monochromatic source overcomes all of these potential problems as only photons that conform to the Bragg equation for the conditions of the monochromator assembly (i.e. Al K$\alpha$) will be diffracted and reach the sample positioned on the Rowland circle.

### 3.2.2 Data Acquisition

#### 3.2.2.1 Core Level Spectra

The survey spectrum in XPS or AES will generally be followed up by the acquisition of spectra around the elemental peaks of interest. As both XPS and AES spectra contain valuable chemical state information, these regions will be recorded at a higher resolution. At this point it is appropriate to consider the factors that influence the physical width of the Auger or photoelectron features. As the Auger process is a de-excitation process involving three electrons, any uncertainty in the energies of the electrons involved will be observed in the resultant spectrum. Thus, Auger transitions involving electrons from the band structure of the atom will be broad. Indeed, for the best quality of chemical state information in AES, the transitions of choice are those involving three core electrons (CCC transitions). That said, excellent chemical state information can be obtained from some CVV Auger electrons such as the Cu $L_3VV$. Thus, the natural width of Auger transitions is dominated by the process itself rather than instrumental factors.

In XPS, the natural line width is very small, although it can influence the shape of the spectrum when recorded at very high resolution. To obtain high-quality core level spectra, a monochromated Al Kα source is used which provides an X-ray line width of 0.25 eV in the best cases (cf. 0.8 for Mg Kα and 1.0 eV for Al Kα, from non-monochromated sources). The narrowness of the monochromated source is combined with a very high-resolution mode for the concentric hemispherical analyser (CAE mode at, typically, 5 or 10 eV pass energy). This enables fine structure in the core-level spectra to be seen and the widespread use of monochromatic sources in the early 1990s led to a step change in the level of information attainable from core-level spectra in XPS, particularly for the analysis of polymers. The complexity of core-level XPS spectra recorded at high resolution has seen a parallel growth in the level of sophistication of peak fitting routines to enable the analyst to assign the various components of a convoluted spectrum with a high degree of confidence. Important parameters in such an analysis include:

- the number of components
- the shape of the peak (usually a Voight function – a combination of Gaussian and Lorentzian distributions)
- degree of asymmetry (important in metals which are asymmetric as a result of core-hole lifetime effects)
- the width of the components (constrained together or allowed to vary)
- the shape of the background which will be influenced by both elastic and inelastic scattering of the electrons.

The source and use of chemical state information in both XPS and AES is discussed in detail in Section 3.3.

#### 3.2.2.2 Valence Band Spectra

The above discussion relates to core-level XPS peaks but important information is contained in the valence band region which, in practice, extends from the Fermi level up to a binding energy of about 30 eV. This region of the XPS spectrum is very weak, as can be seen from any survey spectrum (e.g. Figure 3.1c). Nevertheless, modern instruments, using high-intensity monochromatic sources and high-transmission analysers, can extract very useful information from the valence band.

The valence band region of an XPS spectrum may be used as a fingerprint, because samples that give very similar, or identical, core-level spectra can be differentiated by examination of the valence band. Figure 3.2a shows the C 1s region of spectra from polyethylene (PE), polypropylene (PP) and an unknown sample. As can be seen, these spectra are indistinguishable. Conversely, the valence band spectra, Figure 3.2b, show clear differences. The two reference

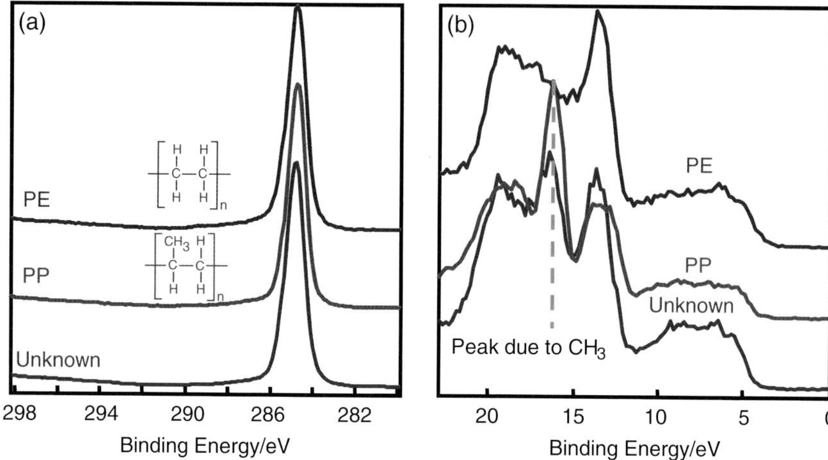


Figure 3.2 A comparison of the C 1s core level (a) and valence band spectra (b) of polyolefin samples.

spectra can then be used in a least squares fitting routine to establish that the PE:PP ratio is approximately 2 : 1 in the unknown sample.

More intense valence band spectra may be obtained using ultraviolet photoelectron spectroscopy (UPS). This uses lower energy photons such as HeI, HeII, NeI, or NeII (with photon energies of 21.2, 40.8, 16.8, 26.9 eV respectively) which provide line widths in the region of 100 meV. Valence band spectroscopy does not, however, provide direct elemental information.

## 3.3 Chemical State Information

### 3.3.1 X-ray Photoelectron Spectroscopy

Long before XPS had developed into a commercially available method for surface analysis it was clear that the spectra produced as a consequence of X-irradiation exhibited small changes in electron energy that were a result of the chemical environment of the emitting atom, ion, or molecule. The XPS chemical shift is the cornerstone of the technique and the reason why high-resolution analysers and accurate calibration of energy scales were seen in XPS long before it was considered a necessity in AES.

Almost all elements in the periodic table exhibit a chemical shift, which can vary from a fraction of an electron volt up to several eVs. Set alongside the line width of the X-ray sources used in XPS (0.25–1.0 eV) it is clear why some form of data processing is often required to extract the maximum level of information from a spectrum. The computer curve fitting of high-resolution XPS spectra is now a routine undertaking, as indicated above, and international standards are now being drafted to provide a unified framework within which such procedures can be undertaken.

The shifts observed in XPS have their origin in either initial state or final state effects. In the case of initial state effects, it is the charge on the atom prior to photoemission that plays the major role in the determination of the magnitude of the chemical shift. For example, the C—O bond in an organic polymer is shifted 1.6 eV relative to the unfunctionalised (methylene) carbon, whilst C=O and O-C-O are both shifted by 2.9 eV. In essence, the more bonds with electronegative atoms that are in place, the greater the positive XPS chemical shift. This is illustrated in a striking manner for fluoro-carbon species, see Figure 3.3, the C 1s chemical shifts being larger than those of carbon–oxygen compounds as fluorine is a more electronegative element. For this particular compound the C—F group is shifted by 2.9 eV whilst $CF_2$ and $CF_3$ functionalities are shifted by 5.9 and 7.7 eV respectively. Unfortunately, such examples of the chemical shift are unusually large and, in general, values of 1–3 eV are encountered.

An example of the manner in which the peak fitting of a complex C 1s spectrum is achieved is shown in Figure 3.4. The sample is an organic molecule, the diglycidyl ether of bisphenol A, which is a precursor to many thermosetting paints, adhesives and matrices for composite materials. By consideration of the structure of the molecule, it is possible to build up a synthesised spectrum, the relative intensities of the individual components reflecting the stoichiometry

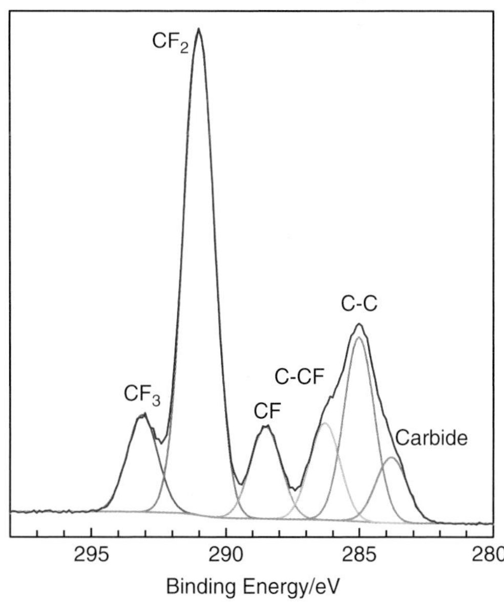


Figure 3.3 C 1s spectrum from a fluorocarbon material.

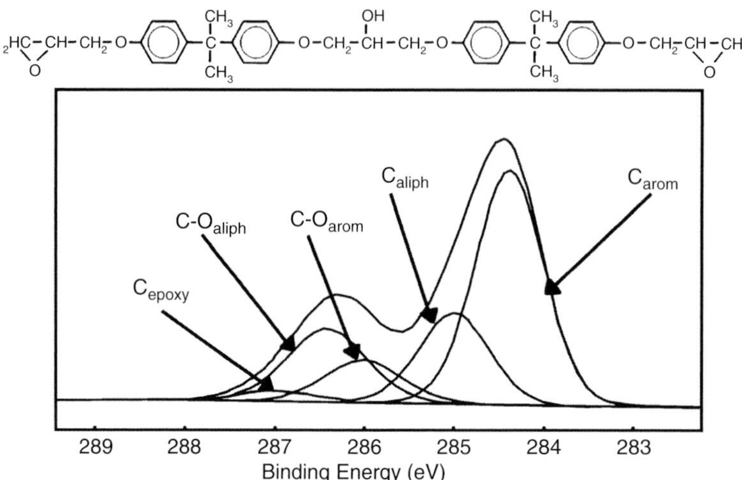


Figure 3.4 C 1s spectrum of the basic building block of epoxy products, the diglycidyl ether of bisphenol A, the structure of which is shown above the spectrum. This spectrum was recorded using monochromatic Al Kα radiation.

of the sample. For polymer XPS at this level of sophistication the resolution attainable with a monochromatic source is essential.

Final state effects that occur following photoelectron emission, such as core-hole screening, relaxation of electron orbitals and the polarisation of surrounding ions are often dominant in influencing the magnitude of the chemical shift. In most metals there is a positive shift between the elemental form and mono-, di- or trivalent ions but in the case of cerium the very large final state effects give rise to a negative chemical shift of ~2 eV between Ce and $CeO_2$. This is, however, an exception and most elements behave in a largely predictable manner.

There are various compilations of binding energies and the most extensive is that promulgated by NIST which is available free of charge over the internet (https://srdata.nist.gov/xps/Default.aspx). This provides a ready source of standard data with which the individual components of a spectrum can be assigned with a high degree of confidence.

One problem that exists with such compilations of data is that many of the values have been recorded from samples that are electrical insulators, which presupposes that the data recorded has then been referenced back to a correctly calibrated energy scale. Such charge referencing is necessary in the analysis of insulators, as the emission of photoelectrons gives the surface a small positive charge, proportional to the photoelectron cross section of the elements contained in the sample. This positive charge has the effect of retarding the outgoing electrons by a small amount thus reducing their kinetic energy which in XPS manifests itself as an increase in the apparent binding energy.

The charge referencing process is the correction of this apparent value back to the 'correct' one.

There are various ways of doing this and one which was very popular in the early days of XPS was the use of the carbon 1s peak from adventitious hydrocarbon as a ubiquitous calibrant. A standard value was assumed for the binding energy of the C 1s line and all other spectral lines corrected by an equivalent amount. Such a facile approach is extremely attractive and was widely used, particularly with non-monochromated XPS where the so-called charging shift was often no more than 5 eV and uniform across the spectrum. The reason for this was that the aluminium window placed between X-ray source and sample generated electrons as the photons passed through it and partially negated the build-up of positive charge at the sample surface, although this would change with the X-ray power used. The major problem was that the C 1s binding energy did not have a 'true' value and assumptions were made about the value to use for the charge referencing process, with energies of 284.6–285.0 being common. Although this approach is still used in certain situations, such as the comparison of a large number of specimens of a similar type, its shortcomings are now widely appreciated. When used, the charge referencing procedure should always be fully described, it is not sufficient to state how the energy scale itself was calibrated (Section 2.5.3). Conversely, when seeking XPS chemical shift data for insulators it is important to pay particular attention to the manner in which charge referencing has been achieved.

With the widespread use of the monochromatic Al Kα source, the use of an external source of electrons became necessary. The most basic form of flood gun is a simple thermionic emitter positioned in line-of-sight of the sample to provide a flux of electrons which would compensate for the build-up of positive charge at the sample surface. Early sources were tuned by the user to the sample characteristics by either minimising the width of a particular peak or correcting (often the C 1s!) to a desired value. This procedure still contained an element of subjectivity but at least the analyst had some control over the manner in which charge compensation was affected. The charge compensation processes on current spectrometers are much improved and are often described as 'turnkey' charge compensation. Some spectrometer designs make use of a magnetic immersion lens around the sample which increases the photoelectron flux, which is particularly useful for imaging XPS systems (although it convolves ARXPS information). Low-energy (flood) electrons are injected into the magnetic field above the sample and spiral down the magnetic field lines, an additional electrode reflects low-energy electron from the sample back to the sample surface. The two processes together form a self-compensation system which ensures the sample surface is always at its natural potential and never over supplied or deficient in electron charge. An alternative approach is to combine flood electrons with a source of low-energy ions, often $Ar^+$, see Figure 6.2 for a schematic diagram of such an arrangement. The combination of the two sources provide a synergistic approach with the ions ensuring the region immediately adjacent to the area of X-ray illumination matches that of the analysis area and thus a more uniform charge compensation by the electron beam is achieved.

Conducting samples such as metals suffer from none of these problems as, being in contact with the spectrometer earth, a ready supply of electrons is available. Problems may arise if a non-conducting oxide is present at the surface or if the samples have been poorly mounted for analysis. Some laboratories routinely check mounting procedures for conducting specimens using a multimeter to ensure continuity between sample and spectrometer ground.

### 3.3.2 Peak Fitting of XPS Spectra

To extract the maximum amount of chemical information from core level XPS spectra, it is often necessary to resort to peak fitting to unravel multiple chemical states of the same element. In certain cases, this is fairly straightforward, but often presents a significant challenge in extracting all the available data. As a starting point for the provision of guidance for peak fitting, the C 1s spectrum of Figure 3.4 will be considered. This is a relatively simple spectrum and one in which a few simple rules of thumb enable peak fitting to be carried out satisfactorily and in a spectroscopically meaningful manner. The latter term is used with care and reflects that all peak fitting routines, whether third party or provided by the instrument manufacturer, are able to treat all parameters (singlet width, intensity, peak position, peak shape for example) as variables, and thus it is straightforward to obtain a fit that is mathematically sound on the basis of a goodness-of-fit parameter. This does not mean that it is spectroscopically sound and the user must constrain some of the variables to ensure the eventual peak fit is meaningful.

The first part of the peak fitting process is to define the window over which peak fitting is to be carried out and select the background shape to be employed. In the C 1s spectrum of Figure 3.4, the peak fitting window is deliberately narrow at 5 eV and excludes the contribution from the $\pi \rightarrow \pi^*$ shake-up satellite. An S-shaped (Shirley) background (see Figure 3.14) has been defined prior to peak fitting although a linear background would have probably been acceptable in this case as the background rise across the peak is small. The first variable that may be considered is the singlet full width at half maximum intensity (FWHM). In the absence of any evidence to the contrary (such as oxide and metal components, see below), it is usual to set all singlets to the same FWHM and the same shape (often a Voigt function comprising of Gaussian peak shape with the addition of up to 30% Lorentzian character, depending on X-ray source and analyser resolution). A good fit between an intense leading (low-binding energy) edge and the proposed singlet can be used to give the operator confidence that the FWHM is reasonable. Other peaks will be positioned at appropriate binding energies and if a structure is known or suspected it is often helpful to constrain these components to the expected binding energy position with a small allowance (say, ±0.2 eV) to allow for fine tuning. Variation of the intensity of each of the components is adjusted on an iterative basis by the peak fitting routine, until the match between the experimental data and the convolution of the individual components is acceptable. Some peak fitting packages provide a graphical output of the differences between experimental and calculated spectra and, although this may provide an indication of where a component may need to be added, many experienced spectroscopists prefer to rely on the 'eye of faith' and experience! In this manner C 1s, O 1s and other core levels from polymeric and organic samples may be fitted quite readily with a high degree of precision.

A frequent requirement is to separate the cationic spectrum of an oxide overlayer from that of the underlying metal, such as the Al 2p spectrum of aluminium foil covered with a thin film of its native oxide, as shown in Figure 3.5.

In this spectrum it is immediately obvious that there are two well-defined components, one representing the oxide at a binding energy of 75.6 eV and a metal component at 72.6 eV. Even before peak fitting it is evident that the metal peak is much narrower than the oxide contribution. The Al 2p peak is an unresolved 2p doublet and thus should be fitted with two, asymmetric, individual components. This has been done for the metallic contribution to the spectrum of Figure 3.4, although the separation of the $2p_{3/2}$ and $2p_{1/2}$ is so small (ca. 0.4 eV) that this is seldom done in practice and a single component is perfectly satisfactory. In the case of the oxide component of the spectrum of Figure 3.5 (75.6 eV) the intrinsic broadness of this component obliterates the spin orbit splitting components and a single component is always used. Once again, the FWHM of the metal and oxide components can be established by achieving a good

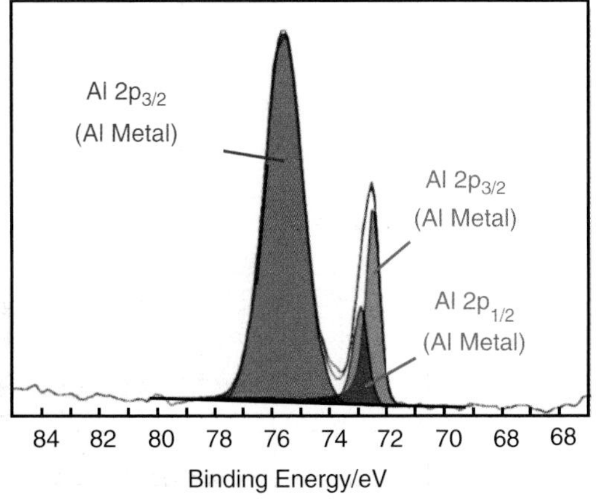


Figure 3.5 Al 2p XPS spectrum of aluminium foil covered with a thin native oxide. Monochromatic Al Kα X-rays were used for the analysis.

alignment of the singlets with the leading and trailing edges of the two components of the experimental spectrum, respectively. A linear background has been used and this can be seen to be perfectly acceptable when the background rise is small. A common requirement when considering spectra of a metal with its native oxide is to report the thickness of the oxide layer. This is done by a simple manipulation of the Beer-Lambert equation, as explained in Section 4.2.1, and the areas of oxide and metal components established by peak fitting can be used for this purpose. It can be seen from Figure 3.5 why it is important to use peak area for this purpose. The intensity of the much broader oxide components would be significantly underestimated if peak heights were used rather than peak areas.

Perhaps the most challenging spectra to handle are those which, unlike the examples of Figures 3.4 and 3.5, are associated with a steeply rising or falling background. There are several reasons for such an occurrence; an element in low concentration may be influenced by the energy loss tail of an adjacent peak from an element at much higher concentration, more often, however the background is seen to rise in a systematic manner across a series of orbitals of the same period. Such a case is the 2p spectra of the 3d transition metals (i.e. chromium to germanium), the background of which shows a maximum in the case of iron. The problem is illustrated in the Fe 2p spectrum of Figure 3.6.

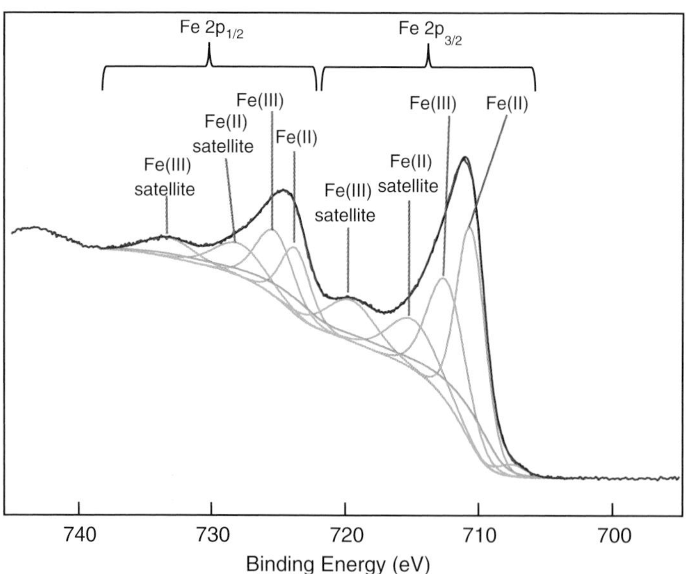


Figure 3.6 The Fe 2p XPS spectrum from a thick oxide layer on air-exposed steel. Peak fitting has been carried out over the entire Fe 2p window following the subtraction of a Shirley background. Components attributable to Fe(II) and Fe(III) and their associated shakeup satellites are included for both Fe $2p_{3/2}$ and Fe $2p_{1/2}$ parts of the spectrum.

The separation of the Fe $2p_{3/2}$ and Fe $2p_{1/2}$ is, of course, significantly greater than that of the Al 2p (13 eV cf. 0.4 eV) so they are clearly separated in the XPS spectra of this region. Although acceptable results can be obtained by simply peak fitting of the $2p_{3/2}$ component, it is good practice, if possible, to peak fit across both the 2p components. There are several reasons for doing this; by constraining separations additional confidence is provided concerning the peak position of the $2p_{3/2}$ singlets and thus the chemistry; similarly the ratio of $2p_{3/2}$:$2p_{1/2}$ should be constant (although not necessarily the expected theoretical value of 2 : 1 as multiplet splitting and the choice of the form of the energy loss background, may distort this); peak widths may vary between metal and oxide, as seen for the Al 2p spectrum of Figure 3.5, but they should remain constant across the spin orbit splitting components. Shake up satellites are an important feature of the 3d transition metals and these are included for both Fe 2p components. All of these features should ensure that the convolution of the individual components matches the experimental spectrum well, across the 40 eV of the peak window. Indeed, the match on the higher binding energy side provides an excellent indication of the fidelity of the peak fitting process.

It is also important to consider background shape. In the example of Figure 3.6 an S-shaped background is used across the entire width of the spectral region. In this manner, the decision has been made to remove a global background prior to peak fitting. This is the only option in the case of many commercial data systems and can work well if the chemical states of the element are distributed homogeneously within the XPS analysis depth. The alternative is to fit a discreet background and tail to each spectral component. This is particularly useful in the case of thin layers, such as corrosion films, where the energy loss background increases as the chemical components considered are positioned at the surface and then towards the bulk of the sample. In such examples, an increasing background intensity needs to be built into the peak fitting strategy, as exemplified in Section 7.3. Notwithstanding this potential shortcoming, there are many examples in the literature where authors have used a global background removal strategy to good effect when carrying out analysis of 2p spectra from samples of this type.

### 3.3.3 Auger Electron Spectroscopy

Historically, electron induced AES is not credited with the ability to yield chemical state information. Early examples of chemical effects in Auger spectroscopy were usually in quasi-atomic spectra excited by X-rays. The reason for this neglect of chemical effects is twofold: the thrust in the early development of AES was the use of analysers, such as the retarding field or the cylindrical mirror analyser, which provided a high level of transmission but at the expense of spectral resolution. Thus, the peaks from early Auger spectrometers were very broad and superimposed on a very intense electron

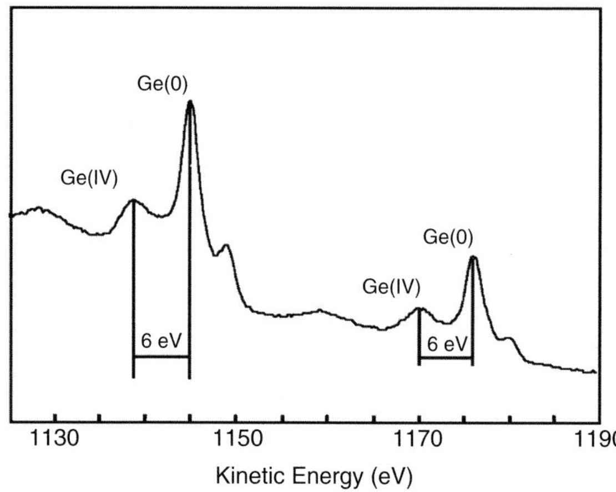


Figure 3.7 Electron excited Auger chemical state information for a germanium single crystal with a thin layer of oxide.

background. This led to the practice of using phase-sensitive detection to acquire differential spectra. Even if there were well-defined, chemical information in the spectra, the practices used for spectral acquisition effectively obliterated it! The other reason is the superposition of the degenerate band structure onto the shape of the Auger peak in the case of *VVV* and *CVV* transitions. This may lead to changes in the shape of Auger transitions from different chemical environments but, generally, not the discrete chemical shift that is observed in XPS core levels. If the two outer electrons are not valence electrons (i.e. *CCC* Auger transitions) a sharp peak may result as observed, for example, in the *KLL* series of peaks of aluminium and silicon, and the *LMM* series of copper, zinc, gallium, germanium, and arsenic. The Ge *LMM* Auger spectra of Figure 3.7 show components attributable to $Ge^0$ and $Ge^{4+}$ separated by over 6 eV.

### 3.3.4 X-AES

The cornerstone of any spectral analysis which relies on peak position to provide information presupposes the ability to determine such values with the necessary accuracy, at least ±0.1 eV in electron spectroscopy. The two possible sources of error are those due to spectrometer calibration and those resulting from electrostatic charging of the sample. The former is easily overcome by accurate calibration of the spectrometer against known (standard) values for copper and gold. The latter is resolved for metallic samples by proper mounting procedures but for insulators and semi-conductors a slight shift as a result of charging is always a possibility. Although it is possible to use an internal standard, such as the adventitious carbon 1s position, this is not particularly accurate and will vary slightly with the form and amount of carbon. A much more attractive method is to make use of the chemical shift on both the Auger and photoelectron peak in an XPS spectrum and to record the separation of the two lines; this quantity is known as the Auger parameter (α) and is numerically defined as the sum of the peaks

$$\alpha = E_{B(i)} + E_{K(ijj)}$$

where $E_B$ is the binding energy of the most intense photoelectron emission peak and $E_K$ is the kinetic energy of the Auger transition. The Auger parameter is numerically equal to the energy difference between the peaks on the binding energy scale. The measured value will thus be independent of any electrostatic charging of the sample. The elements that yield useful Auger parameters in conventional (Al Kα) XPS include F, Na, Cu, Zn, As, Ag, Cd, In, and Te. When using high-energy XPS (discussed in Section 2.3.1) the list can be extended to include Al, Si, P, S, and Cl through to Ti and V. An example of the concept of the Auger parameter using conventional XPS is shown in Figure 3.8, which illustrates the F1s – F *KLL* Auger parameter. As well as providing chemical state information the Auger parameter is, in some cases, able to provide information on crystal structure and relaxation energies. The quantity is more strictly known as the final state Auger parameter and is dominated by the final state relaxation energy. A useful relationship is that the change in Auger parameter relative to the standard state is twice the relaxation energy.

It is also possible to define an initial state Auger parameter (ζ) which requires a knowledge of the kinetic energy of the Auger peak and the binding energies

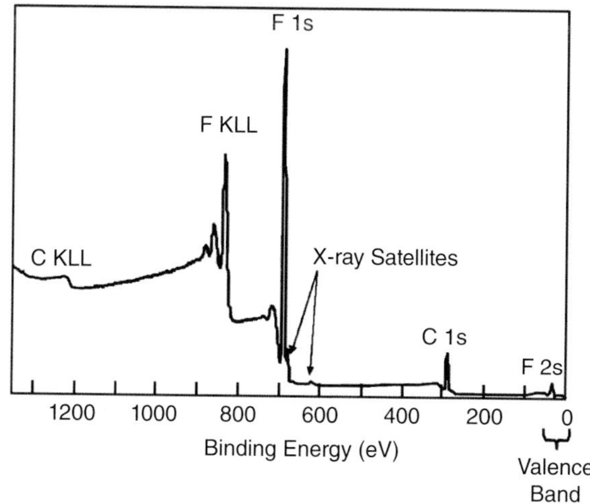


Figure 3.8 Survey XPS spectrum of PTFE showing KLL Auger peaks of fluorine and carbon.

of both the levels involved in the de-excitation process. Thus, for an ijj Auger peak the ζ Auger parameter is defined as:

$$\zeta = E_{k(ijj)} + E_{B(i)} + 2E_{B(j)}$$

This value is less secure than the α parameters described above as it presupposes accurate charge referencing as there is only one kinetic energy term but three binding energy values.

An alternative use of the X-ray induced Auger transition, widely used to elucidate carbon hybridisation is the so-called D-Parameter. Although it is possible to differentiate $sp^2$ and $sp^3$ hybridisation by careful peak fitting of the C 1s spectrum, where the $sp^2$ form will appear to a lower binding energy of some 0.4–0.6 eV relative to the $sp^3$ form, this is not a reliable quantitative approach to determining the relative amounts of the two different hybridisation types. The preferred method is to make use of the C *KLL* Auger transition which is readily extracted from a typical XPS survey spectrum, if recorded at an appropriate resolution. As the Auger transition is a three-electron process, for carbon materials the valence electrons are involved as the internal transition and emitted electrons. This means that the chemical effects present in the valence band electrons are also present (in a heavily convoluted manner) in the C *KLL* transition (where the *L* electrons are of course valence electrons). In order to extract such information a common approach, is to take the first differential of the C *KLL* Auger spectrum and define the D-Parameter. This is the distance (in electron volts) between the maximum and minimum excursions in the C *KLL* Auger spectrum. Such a measurement has been shown to be extremely sensitive to carbon hybridisation, as a result of the influence of the C2p and C2s electrons in the Auger transition. As many authors have shown there is a linear relationship between the value of the D-Parameter and the extent of $sp^2$ and $sp^3$ hybridisation. In practice it is necessary to define the D-Parameter of materials uniquely or predominantly of one hybridisation type; for inorganics this would be graphite ($sp^2$) and diamond ($sp^3$) and for polymers poly(ethylene) ($sp^3$) and poly(styrene) ($sp^2$). The D-Parameter that has been deduced from the C *KLL* is then plotted against the proportion of $sp^2$ character of these standard materials; i.e. 100% for graphite, 75% for poly(styrene) and 0% for diamond and poly(ethylene). The proportion of the $sp^2$ character of the candidate material can then be evaluated by interpolation of the D-Parameter within the reference line described above.

### 3.3.5 Chemical State Plots

It is often convenient to use a graphical representation of the final state Auger parameter (α). As stated above, any electrostatic charging present in the spectrum of insulators will readily be accommodated by the use of the Auger parameter. This is because any reduction in the kinetic energy of the outgoing

Auger electrons is seen as an equal increase in the binding energy of photoelectrons from the same sample. By plotting the kinetic energies of the Auger electrons and the binding energies of the photoelectrons on orthogonal axes it is possible to construct a diagonal grid representing the Auger parameters, equivalent data points along such diagonal lines represent equal values of α. Such a plot is often referred to as a Wagner Plot in recognition of Dr C.D. Wagner who first presented data in this manner. A Wagner Plot for arsenic compounds using As 3d and As *LMM* peaks is shown in Figure 3.9, and it can be seen that the range of As 3d binding energies is smaller (7 eV) than the complementary range of Auger kinetic energies (11 eV), reflecting the larger magnitude of the Auger chemical shift for this element.

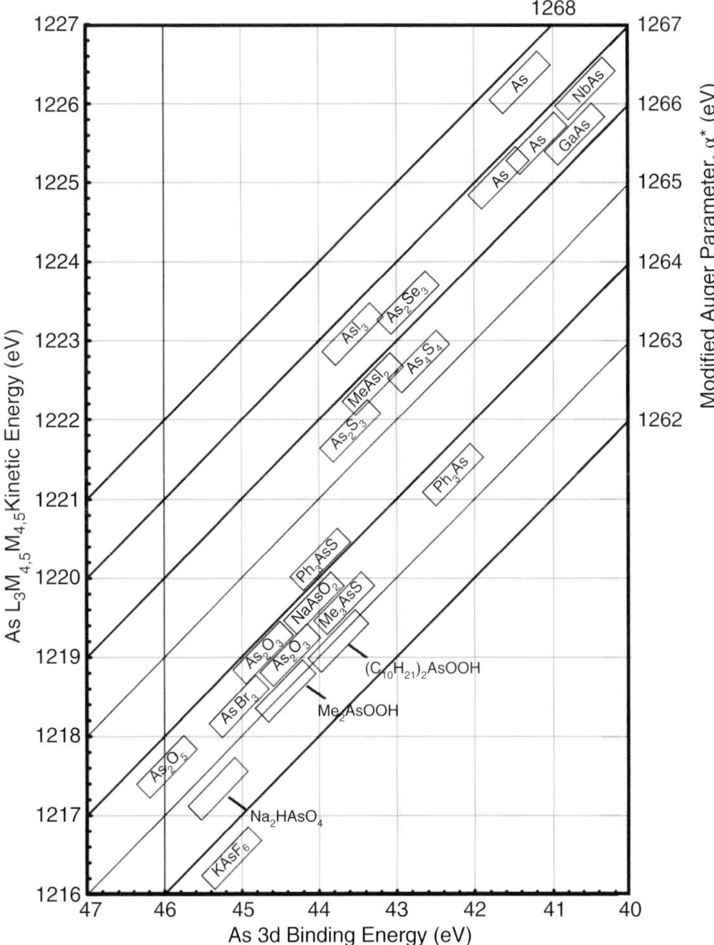


Figure 3.9 Wagner chemical state plot for arsenic compounds.

In cases where an Auger peak is not readily apparent in the spectrum, the bremsstrahlung radiation from a non-monochromated source may be of use. In this manner, it is possible to excite Al *KLL*, Si *KLL*, S *KLL* and Cl *KLL* Auger peaks, for example. In practice this is readily achieved by setting the kinetic energy scale of the spectrometer to the region of interest, the low intensity of these bremsstrahlung-induced features is not a serious drawback, as the electron background in this region of the spectrum (effectively the negative binding energy region of the XPS spectrum) is very low.

Within the high-resolution spectra of individual core levels there may exist fine structure that gives the electron spectroscopist additional information concerning the chemical environment of an atom. The major features in this category are 'shakeup' satellites and multiplet splitting.

### 3.3.6 Shakeup Satellites

Shakeup satellites may occur when the outgoing photoelectron simultaneously interacts with a valence electron and excites it (shakes it up) to a higher energy level; the energy of the core electron is then reduced slightly giving a satellite structure a few electron volts below (but above on a binding energy scale) the core level position. Such features are not very common, the most notable examples being the 2p spectra of the d-band metals and the bonding to anti-bonding transition of the π molecular orbital ($\pi \rightarrow \pi^*$ transition) brought about by C 1s electrons in aromatic organics. The former is best illustrated by the Cu 2p spectrum; a strong shakeup satellite is observed for CuO, as shown in Figure 3.10, but is absent for $Cu_2O$ and metallic copper. In the case of these

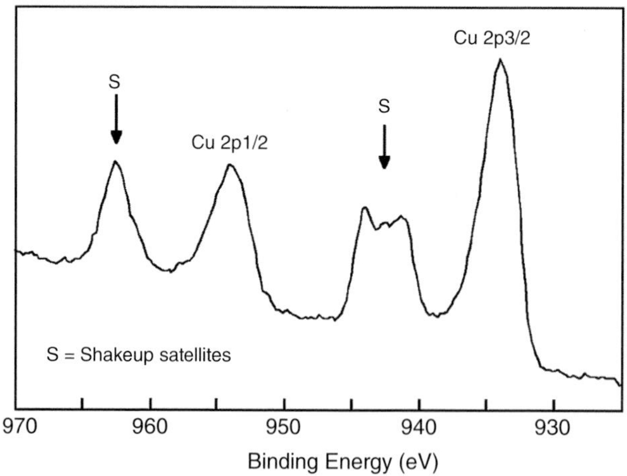


Figure 3.10 Shakeup satellites for Cu 2p spectrum from CuO.

features of inorganic compounds it is now appreciated that the dominant factor in the generation of these satellites is final state effects, such as the screening of the core hole by valence electrons, the relaxation of electron orbitals and the polarisation of surrounding species. An allied feature is the 'shakeoff' satellite where the valence electron is ejected from the ion completely. These are rarely seen as discrete features of the spectrum but more usually as a broadening of the core level peak or contributions to the inelastic background.

### 3.3.7 Multiplet Splitting

Multiplet splitting of a photoelectron peak may occur in a compound that has unpaired electrons in the valence band and arises from different spin distributions on the electrons of the band structure. This results in a doublet of the core level peak being considered. Multiplet splitting effects are observed for Mn, Cr (3s levels), Co, Ni ($2p_{3/2}$ levels), and the 4s levels of the rare earths. The $2p_{3/2}$ spectrum of nickel shows multiplet splitting for NiO, as shown in Figure 3.11, but not for $Ni(OH)_2$ – a feature that has proved very useful in the examination of passive films on nickel.

### 3.3.8 Plasmons

The final type of loss feature to be considered is that of plasmon losses. These occur in both Auger and XPS spectra and are specific to clean metal surfaces. They arise when the outgoing electron excites collective oscillations in the conduction band electrons and thus suffers a discrete energy loss (or several losses in multiples of the characteristic plasmon frequency, about 15 eV for

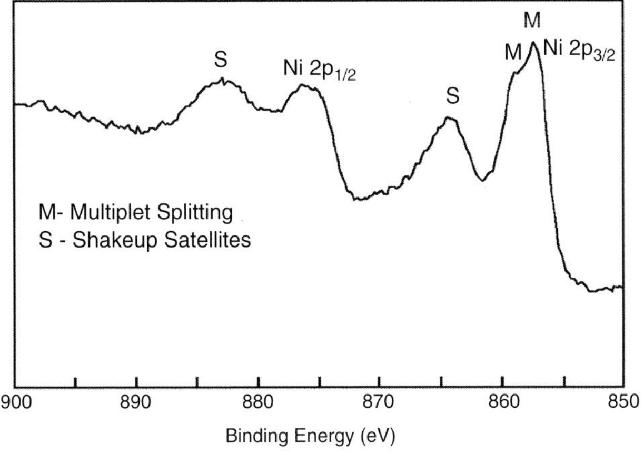


Figure 3.11 Multiplet splitting for the Ni 2p spectrum of NiO.

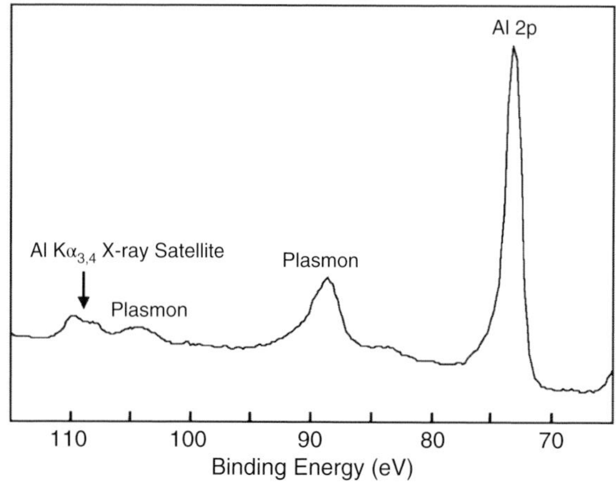


Figure 3.12 Plasmon loss features from clean aluminium.

aluminium). The characteristic plasmon loss peaks for clean aluminium are shown in Figure 3.12. The loss features described above can provide valuable information but some, as in the case of plasmons, merely serve to complicate the spectrum. In either case, it is important to assign them correctly so that all spectral features are accounted for and all elements identified before beginning the calculation for a quantitative surface analysis.

## 3.4 Quantitative Analysis

As both XPS and AES are mature analytical techniques with almost five decades of use in a variety of fundamental and applied scientific investigations, it is to be expected that quantification of the electron spectra produced in the two techniques has reached a high level of precision. This is indeed the case with the most significant body of work in this regard having been undertaken by Seah and co-workers at The National Physical Laboratory, UK. The situation that one is faced with is the need to convert a set of peak intensities to a quantitative analysis, most usually expressed in terms of surface concentration in atomic per cent. Intensities are usually peak areas in XPS and AES (for direct spectra) or peak-to-peak heights (for differential or derivative spectra), representing the elements that have been detected. In principle, this should be a simple undertaking, but the calculation of a quantitative analysis is an undertaking in which the accuracy obtained is in direct proportion to the effort (and consequently the time) that is put into the process. Before reviewing quantification procedures in XPS and AES, it is helpful to consider some of the potential problems facing the analyst in such procedures.

Most algorithms used for quantification assume that the sample is homogeneous within the XPS or AES sampling depth. This may occasionally be the case but in the vast majority of cases the very use of surface analysis methods is because variations at the outer surface are expected, or at least suspected. There are several ways in which such suspicions can be allayed or confirmed including the inspection of the inelastic scattering tail following the peak of interest and comparing intensities of low- and high-kinetic energy peaks (such as Ge3d and Ge2p3/2). This will provide a qualitative appreciation of any structuring that exists within the sampling depth. To provide quantitative measurements of such a hierarchy of layers one must resort to angle resolved measurements or formal analysis of the inelastic background in the manner pioneered by Tougaard, as outlined in Sections 4.2.2 and 4.2.8 respectively. Notwithstanding such a problem, the comparative quantitative analysis can be extremely useful, once one has explained, for example, to a metallurgist that the inclusion of 35 atomic % carbon in the analysis of his stainless steel is not an error but the result of the presence of an adventitious hydrocarbon layer of approximately one nanometre in thickness. In addition to issues associated with the heterogeneity of the sample, there are a number of instrumental and data processing issues that must be addressed, as will be discussed below.

## 3.4.1 Quantification in XPS

### 3.4.1.1 Calculating Atomic Concentration

For many years now XPS has been promoted as a 'quantitative technique' with all the positive accolades that such a title implies. The apparent ease of quantification stems from two important features of the XPS spectrum. Unlike X-ray analysis in the scanning electron microscope (SEM) or AES, matrix effects in XPS are small and, in the vast majority of cases, can be ignored. This means that relative sensitivity factors or photoelectron cross sections can be used to correct a peak intensity into a normalised peak intensity whose value is proportional to the concentration of the element under consideration. The other feature is the inherently low background of an XPS spectrum which is generated, in the main, by the inelastic scattering of the photoelectrons themselves. This means the accurate definition of a peak area is often a quite simple task. Certain elements, most notably the 3d transition metals, have significant backgrounds and, in such cases, the manner of background removal must be carefully considered as mistakes can lead to significant errors in the quantification process.

If one considers the intensity, $I_A$, of a specific XPS transition of element A in a solid from a simplified, first principles approach there are several equations in the literature of the general form:

$$I_A = \frac{\Delta\Omega}{\pi}\int_0^{\infty} J\sigma_A W_A N_a TD \exp[-x/(\lambda\cos\theta)]\, dx \tag{3.1}$$

The terms of Eq. (3.1) are as follows:

$\frac{\Delta\Omega}{\pi}$ is the angular acceptance of the spectrometer

J is the photon flux

$\sigma_A$ is the photoelectron cross section for the element and orbital under consideration

$W_A$ is the angular asymmetry parameter for the element and orbital and the spectrometer employed

$N_A$ is the concentration of atoms A in the solid

x is the depth perpendicular to the surface (integrated to give intensity for the total depth of photoelectron emission)

$\lambda$ is the attenuation length for the electron kinetic energy concerned in the solid being analysed

T is the transmission function of the electron energy analyser

D is the detector efficiency

$\theta$ is the electron take-off angle relative to the sample normal.

Inspection shows that there are essentially three parts to the equation representing the generation of the photoelectrons (J, $\sigma_A$, $N_A$), their transport within the solid into the vacuum chamber of the system (x, $\lambda$, $\theta$), and their analysis and detection within the electron spectrometer itself $\left(\frac{\Delta\Omega}{\pi}, W_A, T, D\right)$.

In practice, this equation is rarely used and the usual procedure is to make a measure of photoelectron intensity for a particular element and core level, correct it for both photoelectron cross section and analyser transmission function and use such a normalised peak area as the basis for a quantitative analysis. Depending on the nature of the cross-section values used (theoretical or experimental) it may also be necessary to consider the electron attenuation length ($\lambda$) explicitly in the calculation. As a starting point for a quantitative analysis, one must be able to measure the photoelectron peak intensity to a suitable level of precision. This is no easy matter but one in which a high level of consistency is required as uncertainties in peak area measurement have a direct, and potentially significant, influence on the accuracy of the quantification process.

The spectrum of Figure 3.13 presents the survey XPS spectrum, acquired using the St Malo Protocol, of a chromium steel corrosion sample.

As the spectrum has been recorded at a 0.4 eV channel width, the peak areas determined are considered to be directly proportional to the concentration of the elements within the XPS analysis volume. All XPS systems are equipped with a data system for spectrum acquisition and processing and this will include

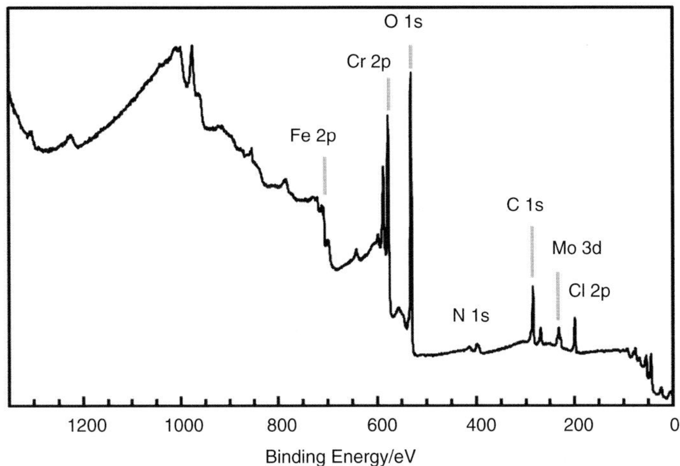


Figure 3.13 XPS survey spectrum of a chromium steel exposed to a corrosive medium. The peaks used for quantification are labelled with element and orbital.

a quantification routine, which will process the peak areas (PA) to provide a quantitative analysis. An algorithm embedded in software will take the area of the most intense photoelectron peak for each element, correct for analyser transmission function and normalise using either experimental or theoretical sensitivity factors (in the case of the latter, a correction for attenuation length is also required) to calculate a normalised peak area (NPA). These NPAs represent the surface concentration of each element. These will normally be summed and the surface concentration reported as an atomic fraction or, more usually, in atomic percent. The process is illustrated below in Table 3.2, for the spectrum of Figure 3.13.

The quantitative data of Table 3.2 assumes a homogeneous distribution of elements within the analysis volume. Inspection of the data and the spectrum of Figure 3.13 shows that this is not the case. The 25% carbon is clearly a surface phase and certainly a contamination layer resulting from atmospheric exposure between corrosion test and spectrometer. Comparison of the energy loss backgrounds of the Cr 2p (575 eV) and Fe 2p (710 eV) peaks indicated that the chromium is a surface phase with the iron buried beneath. Notwithstanding this assumption, made by all standard quantification methods, an important set of data is provided which can form the basis for further exploration of phase distribution, by angle resolved XPS or background analysis if required.

In cases where the sample is known to be homogenous, analysis by XPS indicates the level of precision achieved can be extremely good, and the results are certainly of the order of ±5% of the reported values, achieved by cross-correlation with other analytical methods.

Table 3.2 Typical output for an XPS based quantification.

| Element | Orbital | Peak BE/eV | PA/cps.eV | Atomic % |
|---|---|---|---|---|
| C | C 1s | 284.8 | 765 269 | 25.6 |
| O | O 1s | 530.7 | 3 348 071 | 46.3 |
| Fe | Fe 2p | 707.0 | 582734 | 1.9 |
| Cr | Cr 2p | 576.8 | 4 010 531 | 14.1 |
| Mo | Mo 3d | 231.9 | 451 621. | 1.3 |
| Cl | Cl 2p | 198.8 | 483 264 | 5.6 |
| N | N 1s | 397.8 | 239 856 | 5.2 |

Columns provide information on element orbital used for quantification, binding energy, peak area and the surface concentration in atomic %.

#### 3.4.1.2 Measuring Peak Intensity

Inspection of any XPS survey spectrum clearly shows step-like increases in the background on the higher binding energy side of each photoelectron and Auger electron peak (e.g. the Cu XPS spectra of Figure 2.18). Such an increase is largely associated with the inelastic scattering of the characteristic electrons associated with each transition in the solid material. Inelastic scattering leads to a reduction in the kinetic energy of these outgoing electrons, often referred to as the extrinsic background, which manifests itself in the survey spectrum as the rise in background on the high-binding energy side. The other component of the background in XPS spectra is that which occurs within the binding energy window of the peak itself. This intrinsic background is a result of intra-atomic transitions such as, for example, the shakeup of low energy electrons. The magnitude of the background component of a core-level spectrum varies dramatically across the periodic table with the lightest elements (where the 1s transitions are considered in XPS) having quite modest contributions whilst other elements have a significant contribution from intrinsic and extrinsic losses. If one considers the 2p transitions of the 3d transition metals the background increases and then decreases across the period reaching a maximum at Fe 2p. The removal of a background in such cases requires particular care, as, in order to obtain an accurate and meaningful measure of peak intensity, it is necessary to remove the background contributions to the peak. There are three background constructions that are in common use in XPS; linear, Shirley, and Tougaard. Each will be considered in turn with examples of construction on the Fe 2p spectrum of clean, metallic iron, using integration points at binding energies of 703 and 743 eV, as shown in Figure 3.14. The integrated peak area following the removal of each background type is also included in this figure.

The linear background is the simplest approach to background removal and simply takes the form of a straight-line construction across the spectrum of interest.

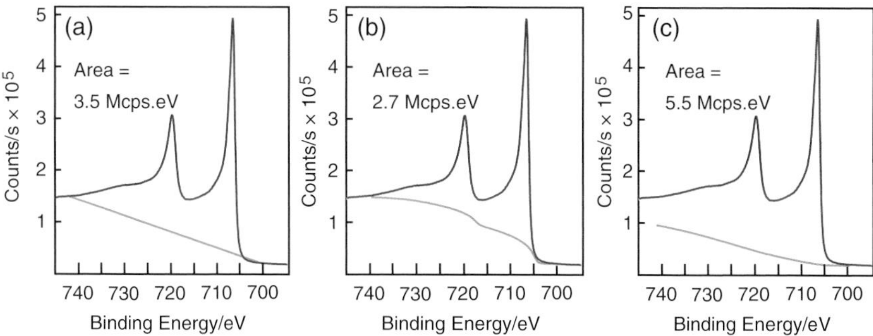


Figure 3.14 Fe 2p spectrum showing the construction of (a) linear, (b) Shirley and (c) Tougaard backgrounds respectively.

This approach might be considered satisfactory where there is an insignificant rise in background but, for all but these cases, the sensitivity to integration points (which are user selected) means that it is extremely difficult to obtain consistent results. Indeed, peak areas may vary by as much as 50% between extreme user-selected windows for removal of a linear background! It should be noted that, when using a linear background, consistency of results is often seen for an individual user, but large variations within user groups. Thus, the choice of limits for the background becomes important, both from a correctness and a consistency viewpoint. Figure 3.14a shows a typical linear background applied to the Fe 2p spectrum, the peak area is 3.5 Mcps.eV. As mentioned above, the linear background is particularly sensitive to the choice of integration points and, if the higher binding energy point is moved 15 eV closer to the Fe 2p, i.e. to 728 eV, the peak area reduces to 2.1 Mcps.eV, a reduction in apparent intensity of some 40%!

The Fe 2p spectrum shows clear asymmetry to the higher binding energy sides of the $2p_{3/2}$ and $2p_{1/2}$ peaks as a result of multiplet splitting and core-hole lifetime effects.

The Shirley background was first proposed in the early days of XPS as a convenient approach to background removal with no theoretical basis (although it is now clear that it accounts for intrinsic losses extremely well). The height of the background at a given point below the peak is a function of the integrated intensity over the peak at kinetic energies above that given point. Background calculation is an iterative process with either the number of iterations being specified by the user or, more usually, calculations continue until variations are very small or negligible. One of the great advantages of this form of background is its relative insensitivity to integration points. The Shirley, or S-shaped, background is illustrated for the same Fe 2p spectrum in Figure 3.14b. A proprietary variation of the Shirley background is the 'Smart' background. This applies a constraint which prevents the background increasing above the raw data, something that sometimes occurs when sets of data (such as from depth profiles or images) are processed automatically. This phenomenon may also occur when a background is applied across a transition showing spin orbit splitting such as the 2p, 3d, or 4f transitions.

The final background form in common usage is that developed by Tougaard. This is based on the modelling of electron scattering in solids and has been developed in the very sophisticated software suite QUASES for the determination of nanostructures in solids. The Tougaard background is unlike the other two in that the background construction extends well beyond the peak(s) of interest as shown in Figure 3.14c, where the 40 eV window is not sufficient to accommodate all of the inelastically scattered Fe2p electrons. Although the Tougaard background is the most rigorous construction, the need to extend to the higher binding energy side of the peaks can make its application to complex spectra difficult as a result of other photoelectron peaks within the background window. However, the analysis of phase distribution by consideration of the energy loss background provides a powerful approach to understanding near surface concentration gradients, as demonstrated by the QUASES software (www.quases.com), as described in Section 4.2.8.

An additional, although rarely used, measure of peak intensity is to measure the total area of the characteristic peak plus its associated background between two user-defined integration points (binding energies). This is unsatisfactory, as this is not only a measure of the peak intensity itself but includes background contributions from transitions at a lower binding energy.

### 3.4.2 Quantification in AES

Although XPS is regarded as an analytical method where the provision of a quantitative analysis is 'the norm', this is not the case for AES and, unlike XPS, a significant body of AES data in the literature is presented with no attempt at quantification. Unfortunately, where quantification is attempted, an analogous approach to that of XPS, making use of atomic sensitivity factors, is often employed which may lead to widely inaccurate results. The reasons for this are dominated by two factors; the matrix effect and backscattering of the primary electron beam giving rise to multiple Auger events.

However, by defining these additional terms, it is possible to describe an expression for the intensity of a particular Auger electron transition, $I_A$, which is analogous to that for XPS, as described in Eq. (3.1).

$$I_A = \frac{\Delta\Omega}{\pi}\int_0^{\infty} I_P \sigma_A \gamma_A N_a RTD \exp[-x/(\lambda\cos\theta)]\, dx \tag{3.2}$$

The terms of Eq. (3.2) have the same meaning as they have in Eq. (3.1) except for:

$I_P$ is the primary beam current (i.e. the incident electron flux)

$\sigma_A$ is the ionisation cross section for the element and the Auger transition by electrons of the primary beam

$\gamma_A$ is the probability of the Auger transition from the sample concerned (matrix effect)

R is the ionisation cross section by electrons scattered in previous processes (the backscattering factor).

The matrix effect (which is taken into account in the term $\gamma_A$) reflects the non-linear dependence of the intensity of an Auger transition of an element with its concentration in the matrix. In other words, the nature of the compound or alloy containing a particular element, as well as the concentration of that element, will influence the intensity of the Auger transition. This, in effect, means that one needs to know the composition of the sample to be able to analyse it successfully. This is not quite the problem it might appear and if a series of similar samples are to be analysed one can use a series of standard samples to establish the magnitude of such a matrix effect, as will be discussed below. Clearly this is not an option for routine analysis and may well explain why such a large proportion of AES data are presented and discussed in a non-quantitative way.

The backscattering factor (R in Eq. 3.2) is necessary as a non-trivial proportion of Auger electrons generated at or near the surface (and consequently contributing to the eventual Auger electron spectrum) are excited by outgoing primary beam electrons that have undergone inelastic scattering events, but still have sufficient energy to excite further Auger electrons. The value of the backscattering factor which, not unreasonably, increases with primary beam energy and, for the values typically used in AES (>5 keV), will have a value greater than one (usually R = 1–2).

Just as with XPS, there are a number of factors which are constant during an analysis by AES so most instrumental terms are straightforward to deal with. It is terms associated with the sample that require careful consideration. However, prior to an approach to the quantification process itself, one must consider the intensity ($I_A$) that is measured. This depends on the form of the spectrum. In the differential mode the intensity measurement is the peak-to-peak height. For low-resolution spectrometers, this is approximately proportional to peak area; for high-resolution studies, fine structure, which becomes apparent in the spectrum, reduces apparent peak-to-peak height. It is for this reason that the integrated peak area of a direct energy spectrum is often preferred for quantitative AES. The relative peak areas in a spectrum will depend on the primary beam energy used for the analysis and also on the composition of the sample. It is the latter, matrix, effect that has prevented the production of a series of AES sensitivity factors of the type widely used for XPS. Instead, it is necessary to fabricate binary or ternary alloys and compounds of the type under investigation to provide calibration by means of a similar Auger spectrum; however, the sensitivity factors produced have a narrow range of applicability. In this manner, it is possible to determine the concentration of an element of interest ($N_A$) as follows:

$$N_A = I_A / (I_A + F_{AB} I_B + F_{AC} I_C + \ldots) \tag{3.3}$$

F is the sensitivity factor determined from the binary standard such that:

$$F_{AB} = (IA / N_A) / (I_B / N_B) \tag{3.4}$$

Such an approach is time consuming and is only justified if a significant piece of work is to be carried out on the sample types involved. In this respect it has close parallels with dynamic secondary ion mass spectrometry (SIMS) where quantification is routinely carried out for depth profiles such as dopant implants in semiconductors, using standard materials of similar implant density.

Various semi-quantitative methods are employed by laboratories throughout the world which relate a measured Auger electron intensity to that of a standard material under the same experimental conditions and this seems to be a fairly satisfactory approach where the time and expense of producing the relevant standard samples is not warranted.

Although a surface analytical study may be an end in itself, knowledge of the concentration of elements near to the surface is often required. To achieve this, some form of compositional depth profiling is required, either by destructive or non-destructive means and this adds another degree of complexity to the interpretation of the resultant spectra, as we shall see later.

# 4

# Compositional Depth Profiling

## 4.1 Introduction

Although both XPS and AES are essentially methods of *surface* analysis, it is possible to use them to provide compositional information as a function of depth. This can be achieved in four ways:

- By manipulating the Beer-Lambert equation (Section 1.6) either to increase or to decrease the depth of analysis. This may be achieved by changing the geometry of the experiment, or the energy of the emitted electrons. These methods are often described as 'non-destructive' because they do not involve the deliberate removal of material from the sample.
- Analysis of the background of inelastically scattered electrons that is always an integral part of the XPS or AES spectrum. This can be evaluated to provide qualitative or quantitative information regarding the hierarchy of layers and nanostructure of the specimen.
- By removing material from the surface of the sample *in situ* by ion sputtering. Analysis is then alternated with material removal and a compositional depth profile gradually built up.
- By sectioning the sample and then analysing the newly exposed surface. This can be achieved either by using a finely focused ion beam to produce a crater whose side wall is analysed or by removing material mechanically using methods such as angle lapping or ball cratering.

The choice of method depends largely upon the required depth of analysis or the thickness of the layers being analysed, as illustrated in Figure 4.1. The choice of method has some dependence upon the type of material being analysed and the equipment available to the analyst and so there is some overlap of the depth ranges.

*An Introduction to Surface Analysis by XPS and AES*, Second Edition.
John F. Watts and John Wolstenholme.

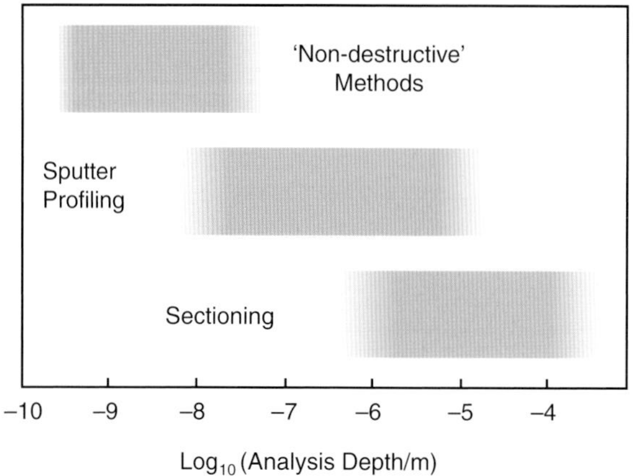


Figure 4.1 The depth range over which each method of profiling is applicable.

## 4.2 Non-destructive Depth Methods

### 4.2.1 Measurements at a Single Emission Angle

The XPS spectrum in Figure 4.2 is the Si 2p region from a smooth flat sample of silicon which has a layer of silicon dioxide (of thickness d) covering its surface. The higher binding energy peak (shown in red) is from the oxidised silicon in the overlayer and the peak at the lower binding energy (shown in blue) is from elemental silicon. The latter appears because the oxide layer is sufficiently thin to allow some fraction of the electrons emitted from the silicon to pass through the overlayer without suffering energy loss. The inset in Figure 4.2 shows the generalised geometry for the collection of photoelectrons, in the case of the spectrum shown the electron emission angle ($\theta$) was 45°.

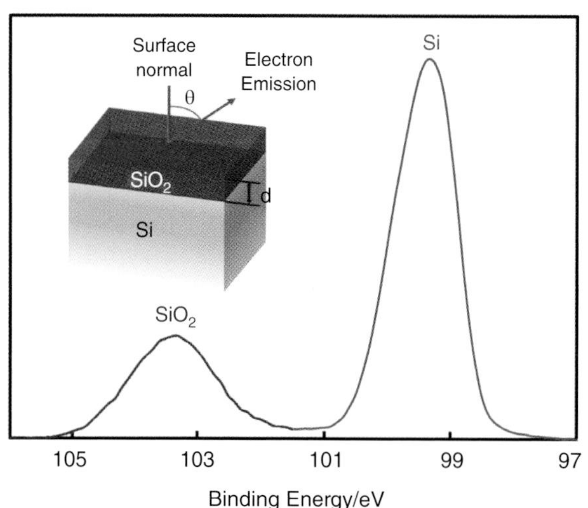


Figure 4.2 The Si 2p region of the XPS spectrum from a silicon sample which has a thin layer of $SiO_2$ on its surface. The inset diagram illustrates the measurement geometry.

By manipulating the Beer-Lambert equation discussed in Section 1.6, and applying it to both the silicon and the silicon dioxide layer, is possible to calculate the thickness, of the overlayer from the ratio of the peak areas.

$$d = \ln\left(1 + R / R_0\right) / \lambda\cos\left(\theta\right)$$

Here, it has been reasonably assumed that the value for the attenuation length, $\lambda$, for the emitted Si 2p electrons is the same regardless of whether they originated in the oxide or in the elemental silicon. Although the atomic density of the oxide will be smaller than that of the elemental substrate, and thus the attenuation length slightly different, all Si 2p photoelectrons will pass through the oxide layer, whether generated there or in the silicon substrate. R is the ratio of the peak intensities ($I_{SiO2}/I_{Si}$) in the spectrum and $R_0$ is the ratio of intensities from a thick layer of oxide and a thick layer of silicon. In the example shown, $\cos(\theta) = 0.7$. Applying this equation, the thickness of the oxide layer was found to be 0.9 nm. The equation becomes a little more complex if widely differing values for $\lambda$ need to be used for the electrons from the substrate and those from the overlayer. This would be the case if the O 1s peak were used to represent the oxide layer instead of the Si 2p peak.

The uncertainty in the thickness measured by this method can be as low as 0.025 nm if the correct procedure is followed, as described in the International Standard ISO 14701.

### 4.2.2 Angle Resolved XPS Measurements

The above calculation of the overlayer thickness contains the implicit assumption that the model chosen for the sample is correct. In this case, a spectrum similar to that shown in Figure 4.2 could be obtained from a thin layer of silicon on a silicon dioxide substrate. If that were the case, the calculation would lead to a silicon thickness of about 3 nm. Indeed, multilayer and other structures can be envisaged that would yield a similar spectrum.

The ambiguity can be avoided by taking measurements at a series of emission angles. This may be done by tilting the sample in the manner illustrated in Figure 4.3 or by collecting angle-resolved data in parallel, as described in Figure 2.28.

The result of taking measurements over an 80° range of angles is shown in Figure 4.4 for a sample of GaAs which has an oxide layer at the surface. It is clear from the montage of As 3d spectra that the oxide peak is dominant at the surface whereas the peak due to As in the form of GaAs dominates at near normal emission angles. This phenomenon is repeated in the gallium spectra (not shown). Note that the spectra in Figure 4.4 have been normalised, the absolute intensity of the spectra decreases as the angle is increased, this is obvious from the way that the noise amplitude increases with increasing angle.

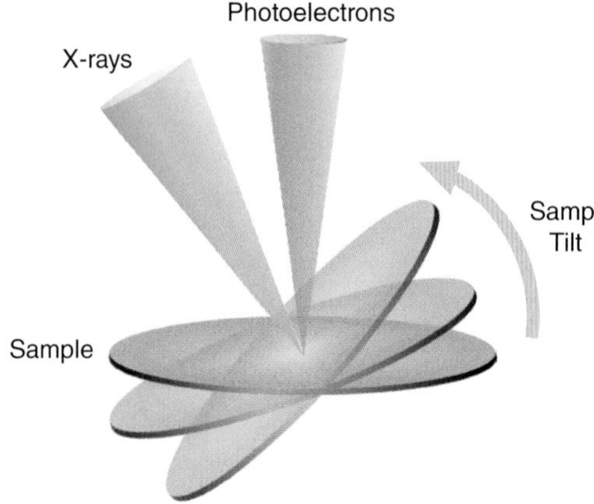


Figure 4.3 Conventional angle resolved XPS (ARXPS), using sample tilting.

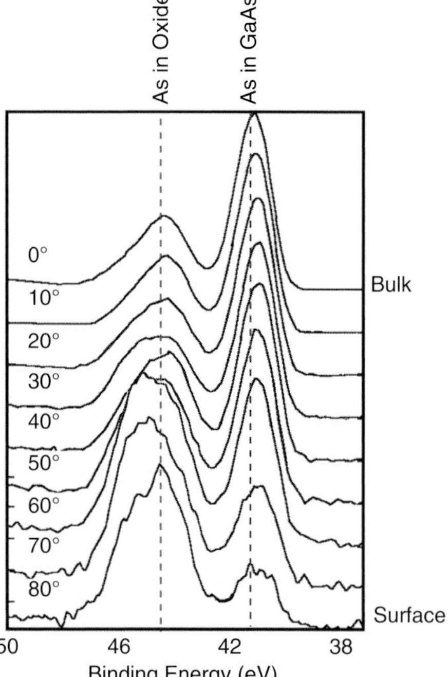


Figure 4.4 Angle resolved XPS (ARXPS) data acquired by tilting the sample. In this case, the sample is gallium arsenide with a thin oxide layer at its surface.

To obtain a qualitative model for the structure of the material, a 'relative depth plot' can be constructed. Such a plot is shown in Figure 4.5a for the oxide layer on gallium arsenide. The plot is constructed for each peak in the spectrum by taking the logarithm of the ratio of the peak area at near grazing emission angle (60°) to that at near normal emission (0°). In Figure 4.5a the plotted points are

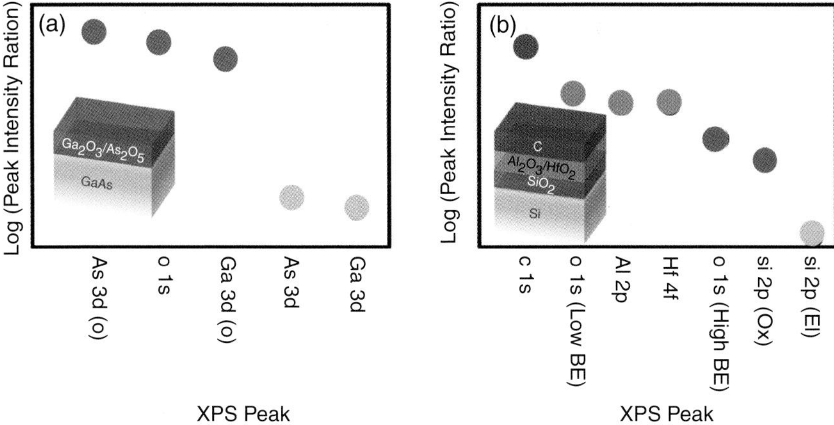


Figure 4.5 Relative depth plots (RDPs) derived from angle resolved spectra from (a) a sample of $Ga_2O_3/As_2O_5$ grown on GaAs; (b) a mixed $Al_2O_3/HfO_2$ grown on a layer of $SiO_2$ on Si. The layer of carbonaceous material is adventitious carbon.

arranged in two groups; one near the top of the plot (the species which are found near the surface of the sample) and one near the bottom of the plot (the species found in the bulk material). This is a quick and simple way to establish the composition and the ordering of the layers in a sample. It provides no information about layer thicknesses.

A relative depth plot (RDP) for a more complex material is shown in Figure 4.5b. This plot shows that there is adventitious carbon at the surface of the sample below which is a layer of mixed oxide ($Al_2O_3$ and $HfO_2$). The fact that the points shown in green on this plot are at approximately the same height suggests that the layer is a mixed oxide, i.e. the two oxides are not present as discrete layers. Below the mixed oxide is a layer of silicon dioxide and below that is the silicon substrate. There were two O 1s peaks in the XPS spectrum from this sample. The RDP confirmed that one of these (the one at the lower binding energy) is associated with the mixed oxide whilst the high binding energy peak is associated with the $SiO_2$.

It is often helpful to construct an RDP before going on to calculate layer thicknesses simply because it unambiguously shows the ordering of the layers and provides a good indication of the chemical composition of each layer. To get an unambiguous representation of the layer structure, each layer must contain at least one element or chemical state of an element which is unique to that layer.

### 4.2.3 Measurement of Overlayer Thickness Using ARXPS

If a set of angle resolved data is collected from samples of silicon each having a layer of silicon dioxide at its surface then, for each sample, $\ln(1 + R/R_0)$ may be plotted against $1/\cos(\theta)$ and a linear plot is expected from each sample (see Section 4.2.1 for the relevant equation and the meaning of each symbol). The gradient of each plot should be equal to the thickness of the oxide layer divided by the attenuation length for Si 2p electrons passing through the $SiO_2$ layer. Knowing the attenuation length, the thickness of the layer is then easily calculated.

Such a plot is shown in Figure 4.6 for a set of six samples of $SiO_2$ on Si. In each case, the calculated thickness of the oxide layer is shown. As can be seen, not all of the data points were included in the calculation of the thickness. The reason for this will be made clear in the Section 4.2.4.

The thickness measurements shown in Figure 4.6 are compared with those obtained from ellipsometry[1] in Figure 4.7. The line in this figure has a gradient approximately equal to one but there is an intercept of about 1 nm on the horizontal axis. This result indicates that there is good agreement between these techniques if it is assumed that there is a layer of adventitious carbon at the surface (observed in the XPS data) which is included in the ellipsometry measurements. This layer must have similar optical properties to a 1 nm layer of $SiO_2$. The adventitious carbon layer is easily distinguished from the $SiO_2$ layer in XPS and can, therefore, be excluded from the $SiO_2$ thickness calculation.

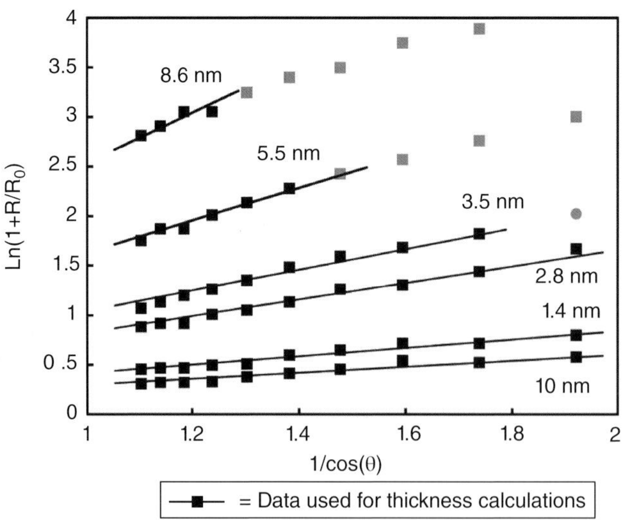


Figure 4.6 Graph showing linearity of the calibration plots for a set of $SiO_2$ on Si samples which have a range of different oxide thicknesses.

1 Ellipsometry is an optical technique, using polarised light, to determine layer thickness with sub-nanometre resolution.

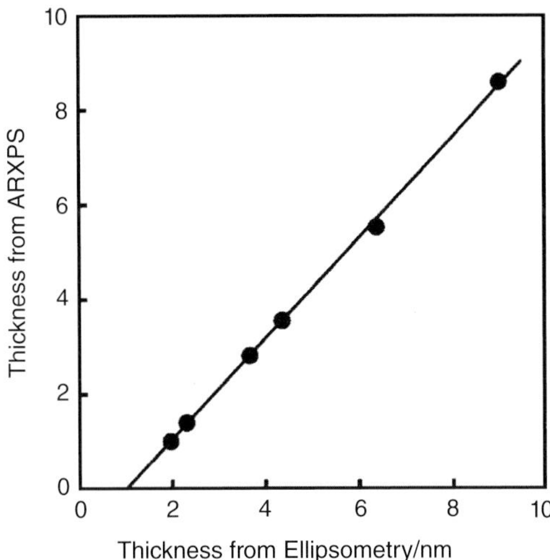


Figure 4.7 A comparison of the thickness of $SiO_2$ layers on silicon measured using angle resolved XPS (ARXPS) with the thickness measured by ellipsometry.

The simple form of the equation for calculating the thickness of layer A on a substrate B may only be used if $\lambda_A \approx \lambda_B$. This is true if the electrons detected from layers A and B have approximately the same energy (e.g. both are emitted from Si 2p). If this is not the case, then a more rigorous equation must be used;

$$R = R_0\left[\exp\left(\frac{d}{\lambda_{B,A}\cos\theta}\right) - \exp\left(\frac{d}{\cos\theta}\left[\frac{1}{\lambda_{B,A}} - \frac{1}{\lambda_{A,A}}\right]\right)\right]$$

In this equation, $\lambda_{B,A}$ is the attenuation length of electrons emitted from material B and travelling through material A. There are, of course, limits to the thickness of an overlayer that can be measured using angle resolved XPS (ARXPS). If the measured thickness of the overlayer is close to the diameter of a single atom then it is likely that the overlayer is incomplete. The lower limit for which the thickness measurement can be used is therefore about 0.2 nm. At the other end of the scale, the signal generated by the substrate becomes weak even when the electrons are collected at near normal emission angles. In general, the thickest layer which can be analysed using this method is in the region of 3λ. For silicon dioxide, this is a little over 10 nm when using the Al Kα X-rays. This upper limit can be extended by using a higher energy X-ray source (e.g. Ag Lα) as discussed in Section 2.3.1.

### 4.2.4 Elastic Scattering

If the data in Figure 4.6 are plotted over a wide angular range then the linearity breaks down, see Figure 4.8. This is a result of the effects of elastic scattering of the electrons which are emitted from deeper regions of the sample. As can

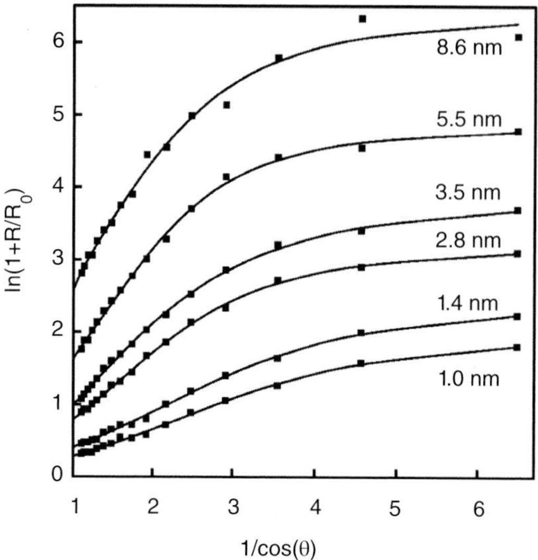


Figure 4.8 The lack of linearity at high-emission angles is due to elastic scattering.

be seen from Figure 1.11 elastic scattering causes electrons emitted from atoms deep within the sample to make a larger than expected contribution to spectra collected at large emission angles.

For this reason, it is usually recommended that spectra collected at emission angles greater than 60° should not be used in thickness calculations (note: $1/\cos(60°) = 2$). However, as can be seen from Figure 4.6, the elastic scattering is affecting the data at smaller emission angles when the layer thickness is large. Conversely, for the thinner layers, linearity breaks down at angles greater than 60°, this is clear in Figure 4.8. For this reason, the figure of 60° should be regarded as a guide. When calculating thickness, the data should be examined before deciding upon the angular range which can be included in the calculation.

### 4.2.5 Multilayer Thickness Calculations Using ARXPS

The calculation of thicknesses for a multilayer sample requires further manipulation of the Beer-Lambert equation. This results in the two equations shown below;

$$\frac{I_i}{I_s} = R_{0,i}\left[1-\exp\left(\frac{-d_i}{\lambda_{i,i}\cos(\theta)}\right)\right]\exp\left[\frac{1}{\cos(\theta)}\left(\sum_{j=1}^{j=n}\frac{d_j}{\lambda_{s,j}} - \sum_{j=1}^{j=i-1}\frac{d_j}{\lambda_{i,j}}\right)\right]$$

$$\frac{I_i}{I_{i+1}} = \frac{R_{0,i}}{R_{0,i+1}}\frac{\left[1-\exp\left(\frac{-d_i}{\lambda_{i,i}\cos(\theta)}\right)\right]}{\left[1-\exp\left(\frac{-d_{i+1}}{\lambda_{i+1,i+1}\cos(\theta)}\right)\right]}\exp\left[\frac{1}{\cos(\theta)}\left(\sum_{j=1}^{j=i}\frac{d_j}{\lambda_{i+1,j}} - \sum_{j=1}^{j=i-1}\frac{d_j}{\lambda_{i,j}}\right)\right]$$

The first of these describes the ratio of the signal intensity from layer i ($I_i$) to that from the substrate ($I_s$). The second equation describes the signal intensity from layer i to that from layer i + 1 ($I_{i+1}$). In these equations, the thickness of layer i is given by $d_i$. $\lambda_{i,j}$ is the attenuation length of electrons emitted from layer i and travelling through layer j. It is not usually possible to assume that the value for $\lambda$ is the same for all emitted electrons in all layers and so the equations cannot be simplified in the way that they were for an oxide layer on its own metal.

The thickness of each layer may be calculated by using least squares fitting of the experimental data, using the values of $d_i$ as fitting parameters. A peak must be chosen for each layer which is unique to that layer.

The use of this method may be illustrated with reference to a set of samples each consisting of a layer of $HfO_2$ deposited on a layer of $SiO_2$ on Si. In each case the $SiO_2$ layer was nominally 1 nm. The $HfO_2$ was deposited using atomic layer deposition (ALD). ALD is a cyclical growth process, the thickness of the layer is dependent upon the number of cycles applied.

Figure 4.9 shows the ARXPS spectra from a sample which has been subjected to 20 ALD cycles. The O 1s peak from $HfO_2$ occurs at a lower binding energy than the O 1s peak from $SiO_2$ which is why the peaks in the oxygen spectrum behave differently with increasing angle. Before measuring the thickness of the layers, it is beneficial to examine the RDP to ensure that it is consistent with the assumed structure for the sample. The RDP constructed from the spectra in Figure 4.9 is shown in Figure 4.10. Data from the C 1s spectrum is included in the RDP to confirm that carbon is present only as a surface contaminant.

The least-squares fitting method, based on the equations above, was applied to a set of similar samples each having been subjected to a different number of ALD cycles. The thickness of the $HfO_2$ and the $SiO_2$ layers may then be plotted against the number of ALD cycles to produce a growth curve, Figure 4.11. Some of the commercially available data processing software packages have the facility to calculate the thickness of layers such as these using the above equations.

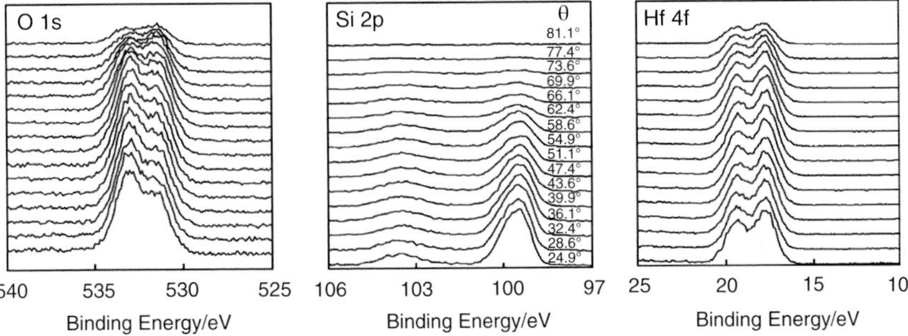


Figure 4.9 O 1s, Si 2p and Hf 4f spectra measured at each of 16 angles from a sample of ~1 nm $SiO_2$ on Si on which $HfO_2$ has been grown using 20 cycles of atomic layer deposition (ALD). The C1s spectra (not shown) were also recorded.

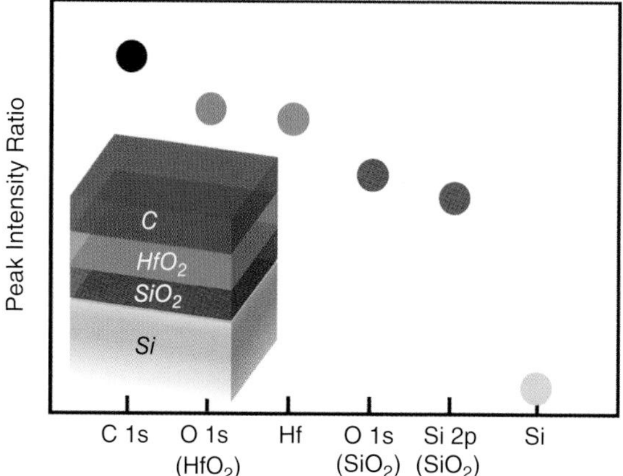


Figure 4.10 The relative depth plot (RDP) derived from the spectra shown in Figure 4.9.

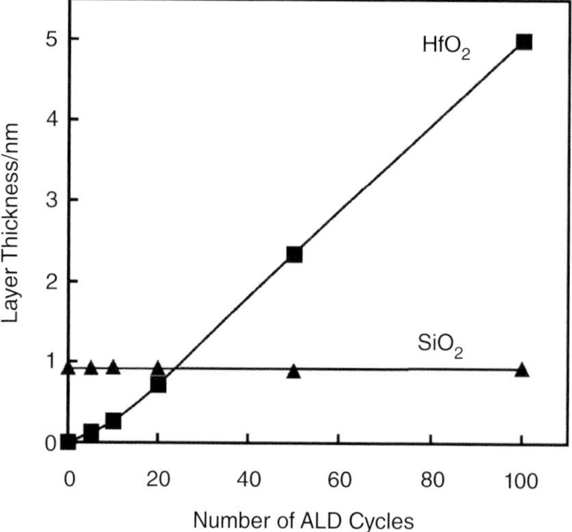


Figure 4.11 The dependence of the thickness of $HfO_2$ upon the number of atomic layer deposition (ALD) cycles used to deposit it on a layer of $SiO_2$ on Si.

Examination of the data shown in Figure 4.11 shows that growth is not linear with ALD cycles and the thickness of some of the $HfO_2$ layers is less than the thickness of a monolayer of the material. These factors suggest that islands of $HfO_2$ are initially formed on the $SiO_2$. Further evidence for this can be found using low energy ion scattering spectroscopy (LEISS), see Figure 4.12a. All samples except for the one which has undergone 100 ALD cycles show the presence of at least some $SiO_2$. As LEISS is sensitive only to the top monolayer of the sample, this can be taken to mean that those samples which show $SiO_2$ to be present in the LEISS spectrum have an incomplete coverage of $HfO_2$. Analysis of the LEISS the data in Figure 4.12a allowed the coverage plot,

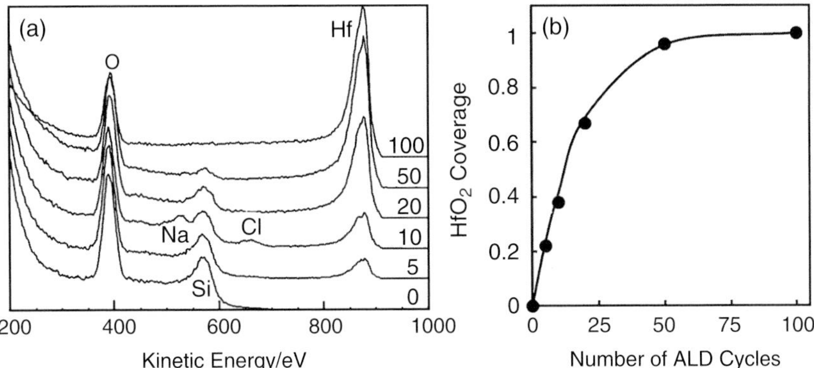


Figure 4.12 (a) The low energy ion scattering spectroscopy (LEISS) spectrum from the samples used to construct the curves in Figure 4.11, the number of atomic layer deposition (ALD) cycles used to grow each layer is shown on the right of each spectrum; (b) the coverage of $HfO_2$ on each sample, derived from the low energy ion scattering spectroscopy (LEISS) spectra.

Figure 4.12b, to be produced. The NaCl present in one of the LEISS spectra is presumably due to surface contamination.

As with the calculation of the thickness of single layers, care must be taken when choosing the range of angles used for a multilayer calculation. Avoid using angles which are significantly affected by elastic scattering.

### 4.2.6 Compositional Depth Profiles from ARXPS Measurements

No unique transformation from angle dependent intensities to depth dependent concentration exists. This implies that a least-squares fit of trial profiles to experimental data is not sufficient to determine accurate concentration profiles. The concept of 'Maximum Entropy'has therefore been introduced to produce a smooth profile, avoiding the over fitting which a method based on least squares fitting would produce.

The method involves the selection of a trial random structure for the sample from which the expected angular data can be calculated. These data are then compared with the experimental data and the misfit calculated. The trial structure is then adjusted with the aim of minimising the misfit. This process is repeated many times (tens of thousands of iterations are commonly used). The process is summarised in the flowchart shown in Figure 4.13. It is recommended that the whole procedure is repeated several times using a different starting trial structure each time. This reduces the risk of the process resulting in a local minimum instead of the 'global' minimum which can be generated from the data.

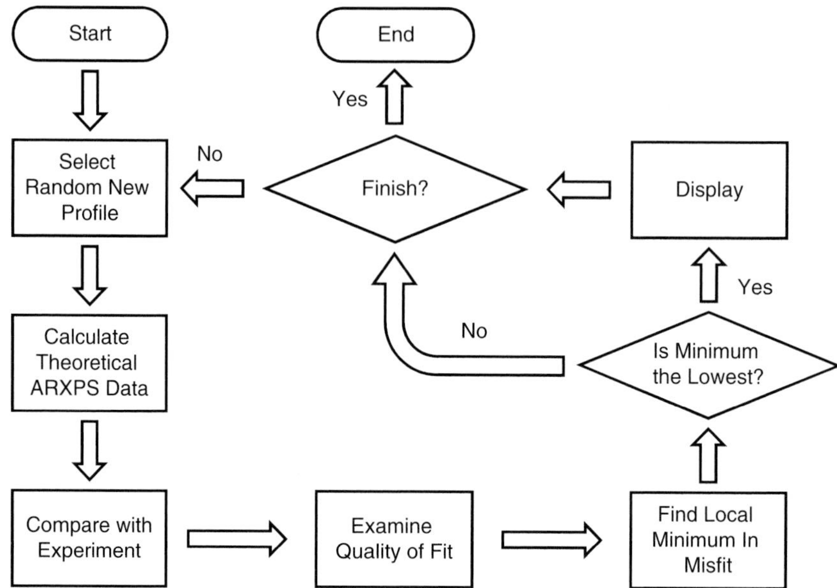


Figure 4.13 Flowchart showing how concentration depth profiles can be generated from angle resolved XPS (ARXPS) data using computational methods such as maximum entropy.

When generating profiles that involve compounds, it is usually better to force the trial structures to adopt the stoichiometry of those compounds. For example, when the sample contains $SiO_2$, the ratio of the Si 2p and O 1s peaks in the trial solutions is forced to be that which would be obtained experimentally from $SiO_2$. If the O 1s peak appears in two materials in the sample (e.g. both $SiO_2$ and $Al_2O_3$) then it can appear in the stoichiometric fit units for each oxide and the O 1s signal will then be attributed to Si and Al in the required proportions.

The mathematical details of this method will not be discussed here. Again, there are ARXPS data processing packages available which include this method to derive depth profiles.

Four examples of depth profiles generated using this method are shown in Figure 4.14. The first, Figure 4.14a, is from a layer of $SiO_2$ on Si. The ratio Si : O in the oxide layer is inevitably 1 : 2 because, instead of fitting the data from these elements individually, the data were fitted assuming the stoichiometry $SiO_2$. This is justified because the peak due to oxidised Si is due to $Si^{4+}$ and the only electronegative element in the XPS spectrum is from oxygen. For clarity, the surface layer of adventitious carbon has been omitted from the profile calculation. This profile serves to show that the method is capable of providing the expected result, but it contains no more information than the simple layer thickness calculation. Figure 4.14b is a profile generated from a layer of $HfO_2$

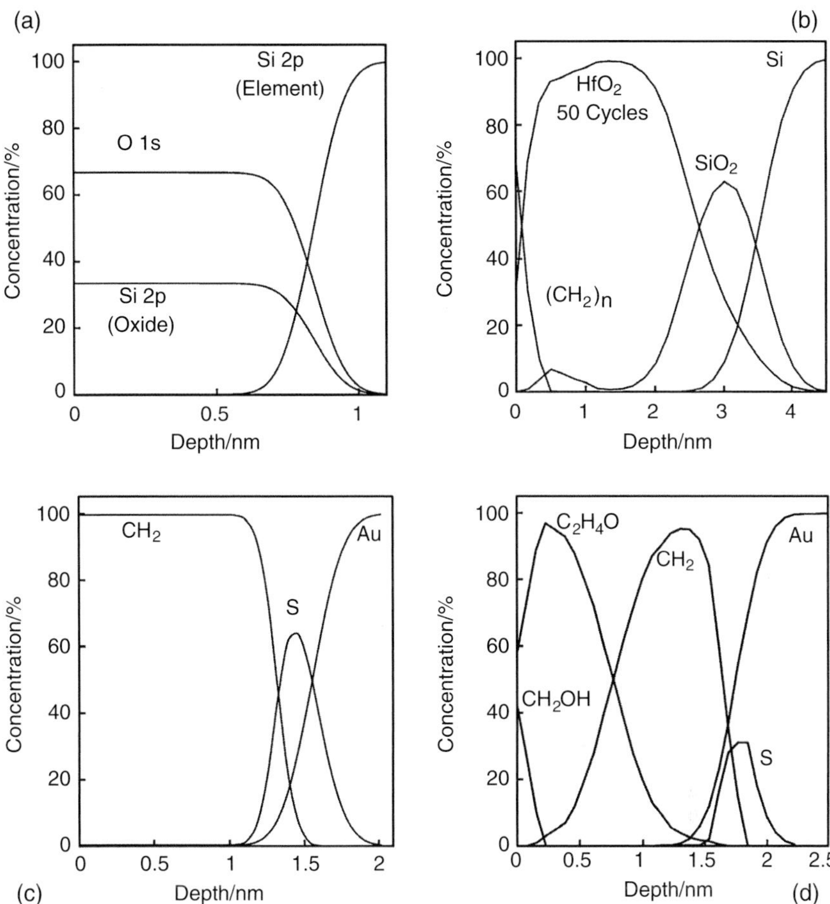


Figure 4.14 Examples of depth profiles generated from angle resolved XPS (ARXPS) data. (a) $SiO_2$ on Si; (b) $HfO_2$ on $SiO_2$ on Si with a layer of carbonaceous contamination at the surface; (c) and (d) self-assembled monolayers (SAM) on gold.

(produced using 50 cycles of the ALD process) on $SiO_2$ on Si. The layer of adventitious carbon has been included in the model, a layer of hydrocarbon has been assumed to have a C:H ratio of 1:2. This profile is not quite what would be expected because it has a small peak in the $SiO_2$ distribution near the surface. This will be discussed further below but a feature such as this would not be observed using the overlayer thickness calculation. Figure 4.14c and d are profiles from two different self-assembled monolayers (SAM) on gold. The first is an alkanethiol (hexadecanethiol) and the second is the more complex 1-mercapto-11-undecyl-tri(ethylene glycol) [S-$(CH_2)_{11}$-O-$(CH_2$-$CH_2$-$O)_3$-H].

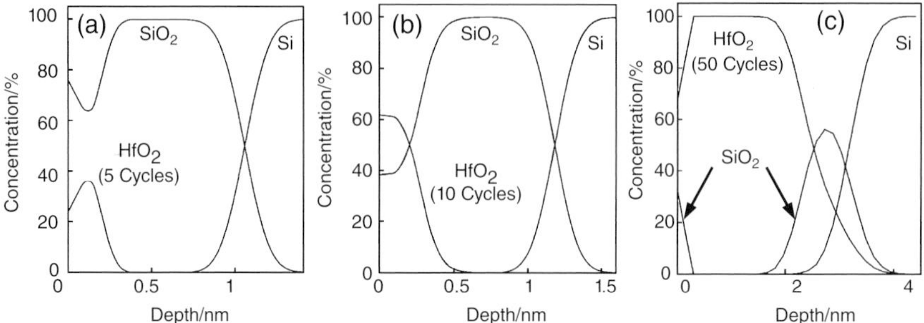


Figure 4.15 Depth profiles from three samples of $SiO_2$ layers on Si which have had $HfO_2$ layers grown on them using atomic layer deposition (ALD). The number of cycles of the ALD process used for each are (a) 5 cycles; (b) 10 cycles; (c) 50 cycles.

These molecules bond to the gold surface via the sulphur atom. This method of profile generation clearly provides a result close to that which would be expected.

When the generated profile does not yield the expected result, as in Figure 4.14b, the reason should be investigated to ensure that it is not an artefact of the method of generation. In the case of the layers of $HfO_2$ produced using ALD, the appearance of $SiO_2$ near or at the surface of the sample is due to the incomplete coverage of the $SiO_2$ by $HfO_2$. Figure 4.15 shows profiles from three of the samples used to construct Figure 4.12. To improve clarity, the carbonaceous layer at the surface has been omitted from the calculation of the profiles. The trend in the quantity of $SiO_2$ appearing at the surface of the profiles in Figure 4.15 is consistent with the trend in $HfO_2$ coverage shown in Figure 4.12.

In this case, the profile calculations do not yield the expected results because they are each a mixture of two profiles, $SiO_2$ on Si and $HfO_2$ on $SiO_2$ on Si.

### 4.2.7 Variation of Analysis Depth with Electron Kinetic Energy

An alternative way of obtaining in-depth information in a non-destructive manner is to examine electrons from different energy levels of the same atom. The attenuation length varies with kinetic energy and, by selecting a pair of electron transitions that are both accessible in XPS but have widely separated energies, it is possible to obtain a degree of depth selectivity. The Ge 3d spectrum (kinetic energy = 1450 eV, $\lambda \sim 2.8$ nm in the oxide overlayer) of Figure 4.16a shows $Ge^0$ and $Ge^{4+}$ components with the oxide component being about 80% of the elemental. In the Ge $2p_{3/2}$ region of the spectrum (kinetic energy = 260 eV, $\lambda \sim 0.8$ nm) the elemental component is much smaller than the $Ge^{4+}$ peak, Figure 4.16b, confirming the presence of the oxide as a surface layer.

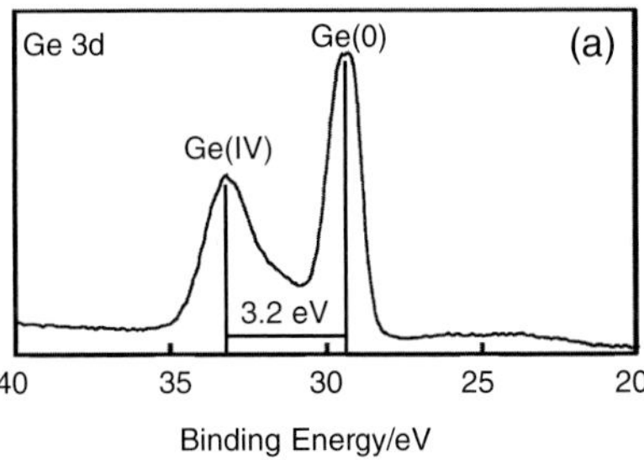


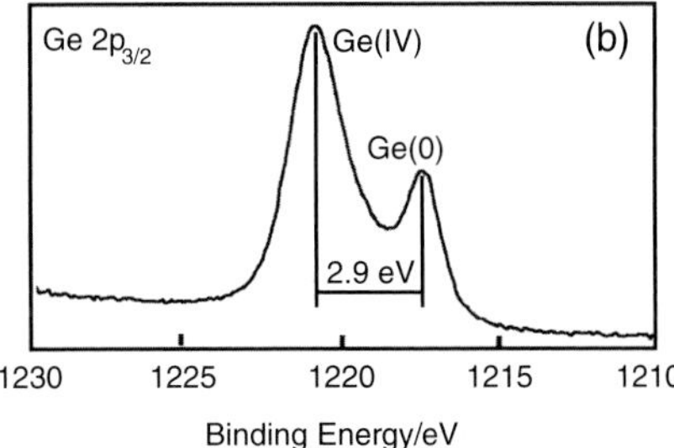


Figure 4.16 (a) the Ge 3d spectrum; (b) the Ge 2p3/2 spectrum from a germanium sample having a thin oxide layer at its surface. This pair of spectra illustrates the variation of sampling depth with electron kinetic energy.

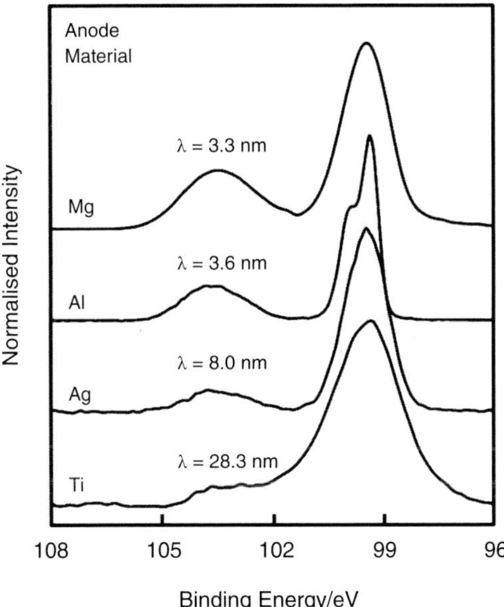


Figure 4.17 Si 2p spectra each acquired using a different anode material in the X-ray source. The spectra are normalised to the elemental silicon peak. The ratio of the oxide peak to the elemental peak decreases with increasing attenuation length.

It is possible to obtain a similar effect by using the same electron energy level but exciting the photoelectrons with a series of different X-ray energies. Figure 4.17 shows four spectra from the Si 2p region from a sample of silicon which has a surface layer of silicon dioxide. Each spectrum was acquired using a different anode to generate the X-rays, see Table 2.1 for the X-ray energy produced by each of the anode materials. The changing intensity ratio $I_{SiO2}/I_{Si}$ with X-ray energy, and therefore electron kinetic energy, clearly shows that the oxide is a surface layer.

The thickness of this oxide layer (1.4 nm) may be calculated from any or all of these spectra (see Section 4.2.1) but for very thick layers (>~25 nm but <~85 nm) thickness could only be measured using X-rays derived from the Ti anode. There are commercially available instruments which can be fitted with a twin anode (Al and Ag) monochromator and a non-monochromated twin anode (Mg and Ti).

### 4.2.8 Background Analysis

The background of inelastically scattered electrons, to the higher binding energy of each photoelectron peak (sometimes referred to as the extrinsic background), contains important information concerning the phase distribution in the near surface region of a sample. The experienced electron spectroscopist will be able to assess this qualitatively by inspection of an XPS survey spectrum. In Section 3.4.1 it was noted that the backgrounds associated with the Cr 2p and Fe 2p peaks of Figure 3.11, had opposing slopes with one element (Cr) being predominantly at the surface of the analysis volume and the other (Fe) being depleted in the surface region. If one considers a substrate buried by a thin overlayer, the peak intensity from the substrate will decrease as the thickness increases, along with a concomitant increase in the background associated with this peak. This arises because, as the overlayer thickness increases, so more electrons from the substrate lose kinetic energy by inelastic scattering within the overlayer, thus reducing the intensity of the peak and increasing the magnitude of the background. At a critical overlayer thickness (which will depend on electron attenuation length), the intensity of characteristic photoelectrons is reduced to a negligible level and the only clue to the substrate composition is an abrupt change in the gradient of the background at a binding energy consistent with that of the substrate photoelectrons.

This phenomenon is clearly seen in the spectrum of Figure 4.18. A thin layer of a commercial silicone-based release agent has been applied to a gold substrate. Peaks from the substrate (Au) and the release agent (C, Si, O) are readily apparent in the spectrum. There are clear features in the background (humps or bulges) associated with the Au 4f and Au 4p doublet pairs. This is a result of the scattering of the electrons emanating from the gold substrate within the organic overlayer.

Quantification of this spectrum, using the approach of Section 3.4.1, would provide an analysis indicative of a layer containing all four elements in proportion to their peak intensity and photoelectron cross section. This is clearly incorrect, and the usual approach is to add a few words to describe the phase distribution, as in Section 3.4.1. This qualitative approach may often lack the

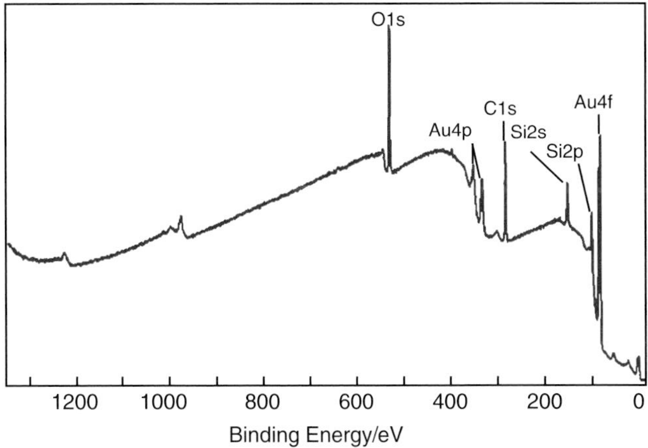


Figure 4.18 XPS survey spectrum of a thin film of silicone release agent on gold.

required rigour and, in such cases, it is usual to resort to a more formal analysis of the background using software such as quantitative analysis of surfaces by electron spectroscopy (QUASES).[2]

QUASES enables the determination of nanometre-scale structures by the analysis of the background in XPS and AES spectra. This approach resolves the problem outlined above but also enables lateral variation in composition to be addressed. Examples of the spectra and background from a number of phase distributions of copper in gold are shown in Figure 4.19, all will yield the same Cu $2p_{3/2}$ intensity, but as can be seen the backgrounds are dramatically different.

The spectrum of Figure 4.19a is representative of an extremely thin Cu film at the surface of gold. This will have virtually no background element as there is little opportunity for the Cu 2p electrons to be inelastically scattered within such a thin layer. Figure 4.19b represents lateral inhomogeneity in the form of a vertical slice of copper to a depth of 5.0 nm within gold. The background will be generated by the slice of copper and, as a function of peak intensity, will be similar to that from a layer of gold of the same thickness, the Cu 2p intensity being much lower. The buried layer of copper (Figure 4.19c) will generate a spectrum with a much higher background as there will be extensive scattering of the Cu 2p electrons within the gold, the background being limited by the number of copper emitters in the thin slice. If there are many

2 www.quases.com

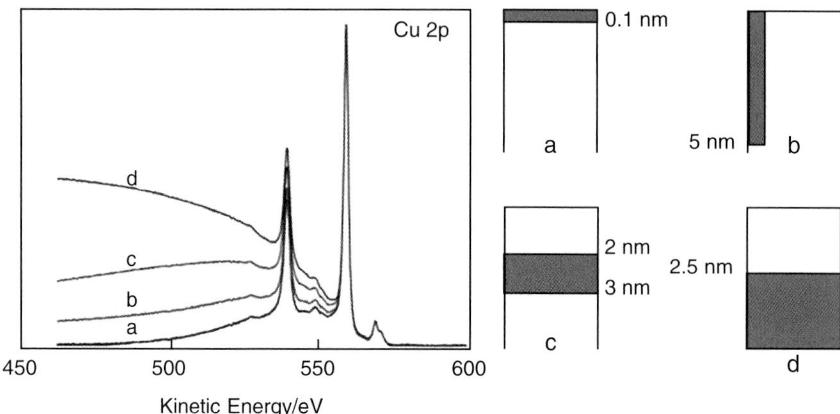


Figure 4.19 The Cu 2p spectra and associated background from four different distributions of copper within a gold matrix. The energy scale is in kinetic energy (binding energy is more usual in XPS) to emphasise the energy loss of the characteristic electrons associated with inelastic scattering. The Cu $2p_{3/2}$ has a binding energy of 932 eV. (*Source:* Courtesy of Professor Sven Tougaard, Physics Department, University of Southern Denmark, Odense, Denmark).

more emitters (in the form of an infinitely thick layer) below a thin layer of gold (Figure 4.19d) the background assumes an opposite gradient as there are many more copper emitters generating electrons which are readily scattered in the gold overlayer. This type of background can often take on a 'bell-shaped' orientation if a sufficiently large energy range is inspected, see Figure 4.18, where as much as two-thirds of the characteristic Au 4f and Au 4p electrons have undergone inelastic scattering in an organic overlayer no more than 3 nm in thickness.

In principle, there are two approaches that can be taken in the use of software that calculates the extrinsic background from a model structure of the specimen:

1) The analysis of the experimental spectrum until all background contributions are accounted for (this might be regarded as a top-down approach).
2) The development of a background that matches the experimental data by additions to a standard spectrum (the bottom-up approach).

In both cases, the crucial step is the postulation of a particular phase distribution of the elements detected (known in QUASES as the nanostructure). By adjustment of this phase distribution it is possible to model the appropriate background for subtraction from the experimental data and comparison with standard data with extrinsic losses removed or addition to standard

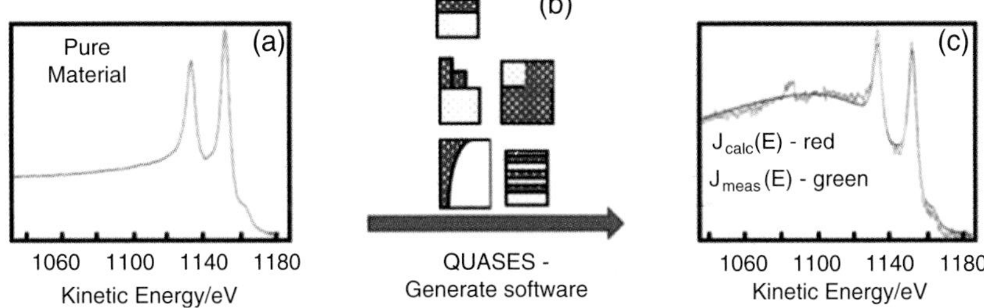


Figure 4.20 (a) The Au 4d spectrum from pure gold; (b) the trial nanostructures used by quantitative analysis of surfaces by electron spectroscopy (QUASES) to calculate the spectrum and background; (c) comparison of the experimental spectrum (green) with the spectrum calculated using QUASES. The procedure is to apply different nanostructures until the calculated spectrum matches the experimental spectrum as indicated in the upper part of the figure. NB: Energy scale is in kinetic energy, the Au $4d_{5/2}$ has a binding energy of 334 eV. (*Source:* Courtesy of Professor Sven Tougaard, Physics Department, University of Southern Denmark, Odense, Denmark).

experimental data, the convolution of the two is then compared with the experimental spectrum. The latter process, using QUASES-generated software is shown schematically in Figure 4.20.

## 4.3 Depth Profiling by Sputtering with Energetic Ions

### 4.3.1 The Sputtering Process

Although the non-destructive methods described above are extremely useful for assessing compositional changes in the near surface region of the material, to obtain data from depths greater than this it is necessary to remove material by ion bombardment within the spectrometer.

The literature available on the subject of ion beam-solid interactions is enormous (see bibliography for examples). All that can be achieved here is to describe the general principles and the possible causes of profile distortion. The primary process is that of sputtering surface atoms to expose underlying atomic layers. At the same time, some of the primary ions are implanted into the substrate and will appear in subsequent spectra. Atomic (cascade) mixing results from the interaction of the primary ion beam with the sample and leads to a degradation of depth resolution. Enhanced diffusion and segregation may also occur and will have the same effect. The sputtering process itself is not straightforward; there may be preferential sputtering of a particular type of ion or atom. Ion-induced reactions may occur; for instance, copper (II) is reduced to copper (I) after exposure to a low-energy, low-dose ion beam. As more and more material is removed so the base of the etch crater increases in roughness and eventually interface definition may become poor.

A high-quality vacuum is important if a good depth profile is to be measured. If there are high partial pressures of reactive impurities present, the surface which is analysed may not reflect the original material composition. This is because sputtering produces a highly reactive surface which can getter residual gases from the vacuum. Oxygen, water, and carbonaceous materials are common contaminants. For this reason also, the gas feed to the ion gun must be free from impurities.

### 4.3.2 Experimental Method

Figure 4.21 shows a flow chart which illustrates the experimental procedure used to obtain a depth profile. The procedure shown in black in this flow chart would result in an XPS depth profile obtained using one type of ion to etch the sample. The procedure for obtaining an Auger profile is entirely analogous. If the analysis is being undertaken using a multi-technique instrument with full computer control, it is possible to use other analytical techniques to obtain complementary profiles. Full computer control of the gas handling system on a

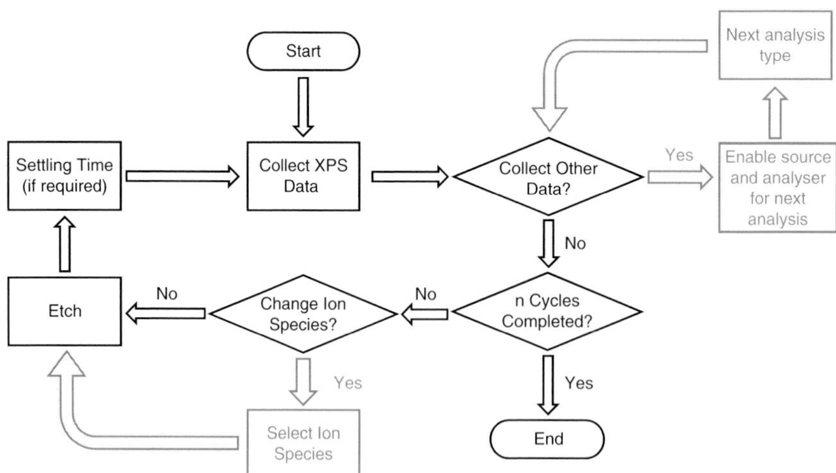


Figure 4.21 Flow diagram showing the experimental method for producing a depth profile in XPS. Following the circuit shown in black would result in an XPS depth profile. The items shown in grey are optional for a suitably equipped instrument.

suitable instrument would allow, for example, ultraviolet photoelectron spectroscopy (UPS) or LEISS profiles to be produced along with the XPS profile. If more than one type of ion source is available, it may be possible to switch ion species mid-profile. It may, for example, be advantageous to switch between monatomic argon ions and argon cluster ions. These analytical options are represented in grey on the flow chart.

The basic profiling experiment usually begins with an analysis of the undisturbed surface using either XPS or AES. The sample then undergoes a period of sputtering (ion etching) using ions whose energy is in the range of a few hundred to a few thousand electron volts. Following this, the ion beam is then switched off (blanked) and the sample is analysed again. This process is continued for a number of cycles, n, until the required depth is reached. When an insulating sample is being profiled, the steady state surface potential during the etching period will be different from that during the analysis. This could cause peak shifts during the early part of the analysis. To overcome this, a settling time can be used during which period the sample is exposed to the analysis conditions but no data are collected. This procedure allows the surface potential to regain its steady state condition before attempting to collect chemical state information.

The following lists possible additional analysis types that might be included with an XPS profile using a suitably equipped instrument:

- AES
- UPS
- Reflection electron energy loss spectroscopy (REELS)
- LEISS
- Energy dispersive X-ray spectroscopy (EDX)
- Raman spectroscopy.

The current density profile of an ion beam is generally not uniform, many approximate to a Gaussian cross section. Such a beam cross section would produce a crater in the sample which does not have a flat bottom. Poor depth resolution would result from extracting data from such a crater. To overcome this difficulty, the ion beam is usually scanned or rastered over an area which is large with respect to the diameter of the beam. Rastering produces a crater which has a flat area at the centre from which compositional data can be obtained.

When collecting the spectroscopic data from the crater, it is important to limit the data acquisition to the appropriate area within the crater, avoiding the area close to the walls where the crater bottom is not flat. This is usually a simple matter in AES because the electron beam used for the analysis usually has a much smaller diameter than the ion beam used for etching the sample. The electron beam can therefore be operated in point analysis

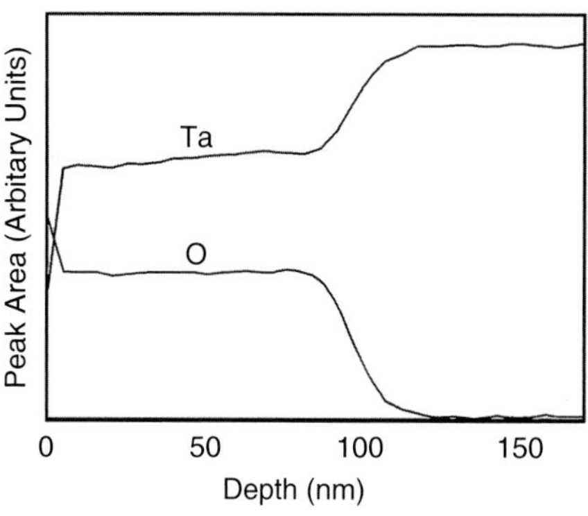


Figure 4.22 XPS depth profile through a layer of tantalum oxide on tantalum metal.

mode in the centre of the crater or rastered over a small area. For XPS, the methods of small area analysis are generally used, as described in Chapter 2 and care must be taken to align the analysed area and the centre of the crater.

Figure 4.22 shows an example of a simple XPS depth profile through a tantalum oxide layer grown on tantalum metal.

There are many points to bear in mind when selecting the experimental conditions for a depth profile. Most of these are concerned with either the speed of acquisition or the depth resolution; the requirements for speed are usually opposite to those for resolution. It is important to understand the principles which govern profiling rate and resolution.

### 4.3.3 The Nature of the Ion Beam

#### 4.3.3.1 Noble Gas Ions

The most commonly used ion beam in electron spectroscopy is one composed of argon ions. Argon is a noble gas and so it does not react chemically with the freshly exposed surface after etching. This is in contrast to secondary ion mass spectrometry (SIMS) where oxygen or caesium are commonly used precisely because these elements react with the sample and increase the secondary ion yield. In some respects, xenon would be a better choice but the differences are marginal and do not, in general, justify its considerably higher cost. Sputtering with noble gas ions is used for both profiling and for gently cleaning the sample surface prior to analysis. The ion gun used for depth profiling may also be used for LEISS measurements which would require the use of helium. Helium, however, is seldom used for depth profiling because of its greater penetration depth and greater propensity to damage the sample. A noble gas ion gun is almost always fitted to XPS and AES spectrometers.

#### 4.3.3.2 Cluster Ions

The use of ion beams consisting of atom clusters has become popular in recent years. These are particularly useful for producing depth profiles of organic or polymeric materials. They profile these materials quickly without seriously damaging the material they leave behind. In general, they profile inorganic materials very slowly. The damaged layer caused by sputtering with monatomic argon ions may be removed using cluster ions.

The first type of cluster ion to be used for profiling was the fullerene, $C_{60}$. XPS instruments having a $C_{60}$ source remain commercially available. A more recent option is to use large argon clusters (containing several thousand argon atoms), these are now becoming very popular. Argon clusters have the advantages that they do not cause contamination of the sample surface or the analysis chamber and the cluster size can be selected to be the optimum for the analysis. A cluster ion source is a very useful addition to an XPS spectrometer especially if the spectrometer is to be used for the analysis of organic materials. There are ion guns available which can be switched between monatomic argon and argon cluster ions during depth profile acquisition.

Figure 4.23 illustrates why the use of argon cluster ions is preferred to the use of monatomic argon ions for the cleaning and profiling of polymers. Figure 4.23a shows the C 1s spectrum from a sample of Kapton[3] in the 'as received' form. The presence of adventitious carbon (surface contamination) masks the chemical states present in the Kapton molecule. Figure 4.23b shows the spectrum which was acquired after a relatively gentle etch with monatomic argon ions. The adventitious carbon has been successfully removed but the relative intensities of the remaining peaks do not reflect the expected structure of Kapton. The peak due to C—C is too large in comparison with the other peaks and the shake-up peaks should be much larger considering the aromatic nature of the Kapton molecule. If the sample is cleaned with argon clusters, however, (Figure 4.23c) the carbon peaks more closely represent the structure of the molecule and strong shake-up features can be seen suggesting that the aromatic nature of the material has been retained.

Similar effects are often seen when sputtering inorganic materials. Figure 4.24 shows three Ta 4f spectra from $Ta_2O_5$, the first (in blue) is the 'as-received'

3 Kapton is a versatile polyimide material developed by DuPont. Its formula is shown in Figure 4.23 and its chemical name is poly (4,4′-oxydiphenylene-pyromellitimide).

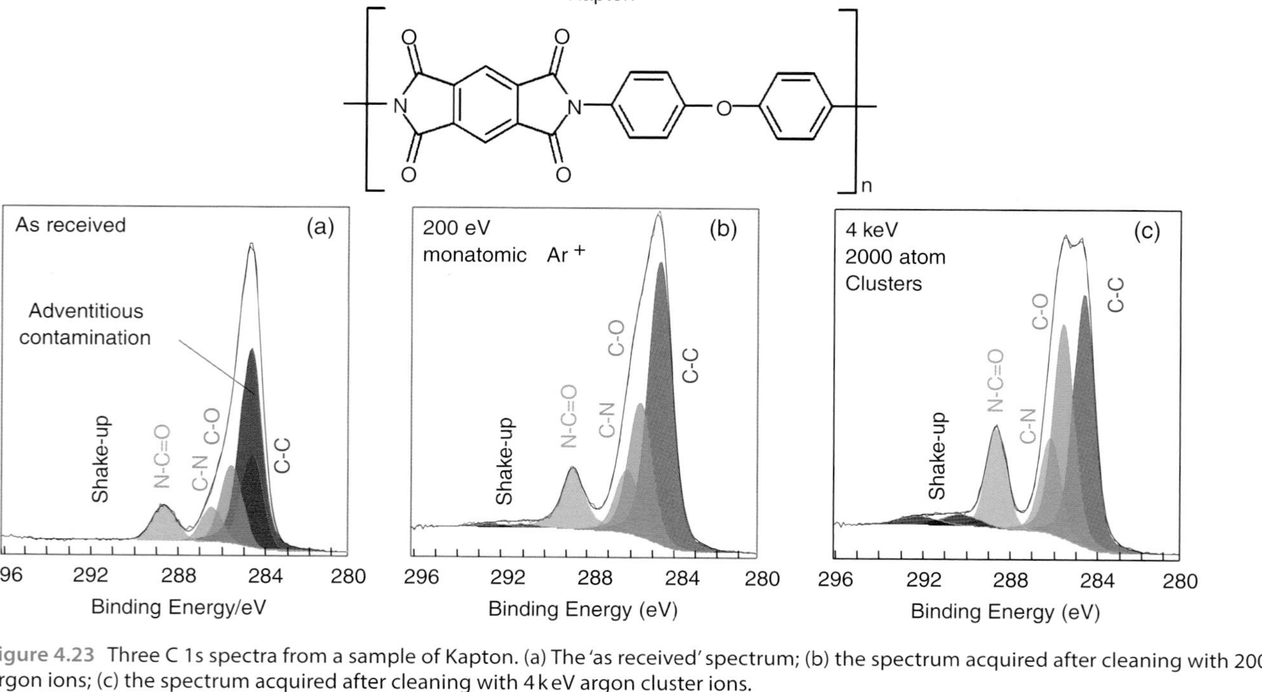


**Figure 4.23** Three C 1s spectra from a sample of Kapton. (a) The 'as received' spectrum; (b) the spectrum acquired after cleaning with 200 eV argon ions; (c) the spectrum acquired after cleaning with 4 keV argon cluster ions.

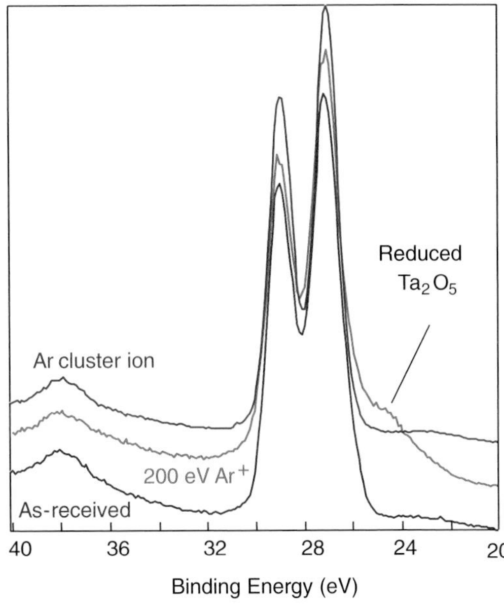


Figure 4.24 Three Ta 4f spectra from a sample of $Ta_2O_5$. The 'as received' spectrum (blue). The spectrum acquired after cleaning with 200 eV argon ions (green). The spectrum acquired after cleaning with 4 keV argon cluster ions (red).

material, the second (green) shows the spectrum after sputter cleaning with 200 eV monatomic argon ions and the third (red) is the spectrum following a clean with argon cluster ions. The green spectrum shows a clear shoulder on the low binding energy side of the peak which is not present in the other two spectra. The shoulder is present because a significant fraction (~30%) of the tantalum has been reduced from the +5 oxidation state by the monatomic argon ions. A result of this phenomenon can also be seen in Figure 4.22 in which there is an initial steep rise in the Ta 4f signal and a corresponding steep fall in the O 1s signal.

#### 4.3.3.3 Metal Ions

These are produced from a liquid metal ion gun (LMIG). The most common metal to be used in this context is gallium. The beam produced by a LMIG may be focused very finely (some may be focused to produce a spot size of <50 nm). In electron spectroscopy, LMIGs are only used in conjunction with AES and the way they are used differs from that shown in Figure 4.21. Typically, these beams are used to produce a crater in the sample having steep, smooth sides, the side wall of the crater is then analysed using AES (either an image or a line-scan). For some analysts, a LMIG is a useful addition to an Auger instrument. This type of analysis is often referred to as focused ion beam (FIB) (focused ion beam) analysis and is commonly used in the semiconductor industry in conjunction with electron microscopy.

### 4.3.4 Sputter Yield and Etch Rate

The sputter yield is the physical property which determines the rate at which material is removed from the sample during bombardment by energetic ions. The definition of sputter yield is:

$$\textit{Sputter Yield}\ (Y) = \textit{Number of atoms removed}/\textit{Number of incident ions}$$

This is the physical property of a sample which has the greatest influence upon the rate at which it can be profiled. To be of practical value, the sputter yield needs to be converted to an etch rate usually given in the units of nm $s^{-1}$.

$$\textit{Rate of removal of surface atoms} = Y \times \textit{Rate of arrival of ions}$$

The rate of arrival of ions is given by:

$$I/e$$

Where $I$ is the ion beam current in amps and e is the charge on the ion in coulombs, we now have:

$$\textit{Rate of removel of surface atoms} = (YI)/e$$

To convert this to a depth scale, we need to know the number of atoms under the beam's rastered area and the atomic layer thickness which can be calculated from the atomic weight and the density of the sample.

$$\textit{The number of atoms cm}^{-3} = (\textit{Target density} \times \textit{Avogadro's number})/\ \textit{Atomic weight} = (\rho N)/w$$

$$(\rho = \textit{density}, N = \textit{Avogadro's number}, \mathrm{w} = \textit{atomic weight})$$

If we make the approximation that the atoms are in a cubic array, the number of atoms in 1 $cm^2$ is $(dN/w)^{2/3}$. Thus, if the beam is rastered over an area of A $cm^2$, the number of surface atoms in the rastered area is $A([dN]/w)^{2/3}$. Similarly, the layer thickness is $[w/(dN)]^{1/3}$. Now, the number of atomic layers removed per second is the number of atoms removed per second divided by the number of atoms at the surface under the ion beam:

$$\textit{Layers per second} = \frac{IY}{e}/\left[A(\rho N/w)^{2/3}\right]$$

To get to the etch rate we have to multiply this expression by the layer thickness:

$$\textit{Etch rate} = \left(\frac{w}{\rho N}\right)^{1/3}\left[\frac{IY}{e}\right]/\left[A(\rho N s w)^{2/3}\right] cm\ s^{-1}$$

This simplifies to:

$$Etch\ rate = IYw / \left( Ae\ \rho\ N \right) cm\ s^{-1}$$

For greater convenience, if the etch rate is expressed in nm $s^{-1}$, the beam current in μA, area in mm and eN as $10^5$ coulombs (a good approximation) then the expression becomes:

$$Etch\ rate = 10^{-2}\ IYw / \left( \rho A \right)\ nm\ s^{-1}$$

### 4.3.5 Factors Affecting the Etch Rate

There are many factors that affect etch rate, some of which are dependent on the material being analysed. Other factors are instrumental and under the control of the analyst. There is the obvious desire for speed in the analysis but, in general, any factor that increases the speed of profile acquisition is likely to have a negative effect upon the depth resolution, as will be seen in the next section. Deciding on the conditions for a depth profile measurement generally requires a compromise.

#### 4.3.5.1 Material

The sputter rate depends upon the chemical nature of the material, not only the elements present but also their chemical state. It is difficult to predict the sputter yield for a material but there are a number of computer simulations available. Some of these can predict sputter yields of elements with reasonable accuracy but they become less reliable when applied to compounds or alloys. It is usually preferable to measure the sputter yield experimentally under the conditions normally employed.

#### 4.3.5.2 Ion Current

As shown above, the etch rate is directly proportional to the ion current. It should therefore be possible to increase the etch rate by using the maximum beam current available. However, in a normal ion gun, the spot size of the beam increases with increasing beam current and so the rastered area must be increased to ensure that the crater bottom remains flat. With this in mind, simply increasing the beam current will not necessarily increase the etch rate. It is the ion beam current density which is the important parameter rather than merely the current.

#### 4.3.5.3 Ion Energy

In the energy range normally employed in XPS and Auger profiling the sputter yield increases with ion energy. At high energy, the sputter yield will reach a maximum. Higher energies also mean smaller spot sizes at a given beam current and so will lead to better crater quality. The higher etch rate will, however, be accompanied by poorer depth resolution because the ions can penetrate deeper into the material causing atomic mixing.

#### 4.3.5.4 Nature of the Ion Beam

In XPS and Auger profiling, it is common to use the ions of the noble gases for sputtering. In the energy range which is normally used, sputter yield increases with increasing atomic mass; xenon ions provide higher etch rates than argon and helium ions provide very much lower sputter yields than argon. In addition, the larger ions penetrate a shorter distance into the material and therefore allow better depth resolutions to be obtained. As previously stated, the cost of using xenon outweighs the benefits it provides.

#### 4.3.5.5 Angle of Incidence

As the angle of incidence (measured from the sample normal) is increased, the sputter rate increases reaching a maximum at about 60°. Above this angle the sputter yield decreases rapidly. The way in which the sputter yield varies with angle is difficult to predict because it depends upon the material being sputtered and the nature of the ion beam. However, since an angle of 60° provides high sputter yields and good depth resolution, many commercial instruments are designed with the ion incidence angle close to this angle.

### 4.3.6 Factors Affecting the Depth Resolution

Depth resolution in a depth profile is a measure of the broadening of an abrupt interface brought about by physical or instrumental effects. A commonly accepted method for measuring depth resolution is to measure the depth range, $\Delta z$, over which the measured concentration changes from 16 to 84% of its total change whilst profiling through an abrupt concentration change, see Figure 4.25.

Many of the factors which affect sputter rate also determine the depth resolution available. Some of these factors relate to the characteristics of the sample, some to the instrument and some the physical process of sputtering.

#### 4.3.6.1 Ion Beam Characteristics

The extent to which the ion beam characteristics affect depth resolution generally relate to the depth range of the ions after striking the sample. This is because the passage of an energetic ion through a solid causes atomic mixing along the whole trajectory of the ion. Hence low ion energy, grazing incidence angles and heavy ions lead to the best depth resolution because these minimise the depth range over which mixing can occur.

Figure 4.25 Definition of depth resolution.

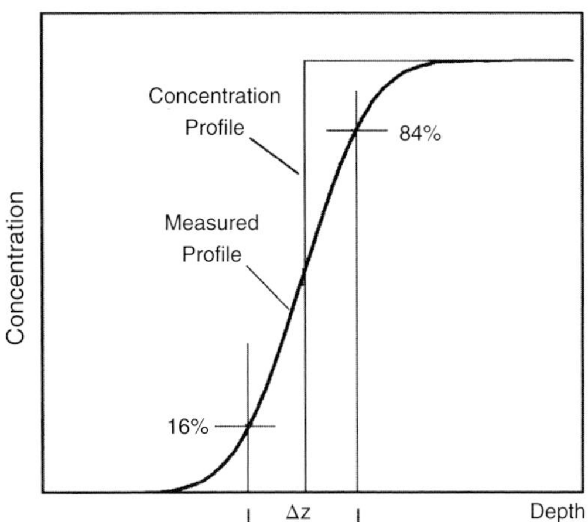


When a cluster ion strikes a surface, it immediately breaks up into its component atoms. Each individual atom then has a very small kinetic energy and is therefore unable to penetrate and disrupt the sample. Under the conditions typically used for profiling with argon clusters the energy per atom is usually in the range 1–10 eV, compared with 100s or 1000s eV/atom for atomic ion bombardment (eg $Ar^+$).

#### 4.3.6.2 Crater Quality

The crater must be as flat as possible over the analysis area. If this is not the case, then information is being collected from a range of depths and resolution suffers. Generally, the raster dimensions should be at least 5 ion beam diameters to get good flatness over a reasonable distance within the crater.

#### 4.3.6.3 Beam Impurities

Chemical impurities in the beam can be minimised by using a high purity gas feed. Other impurities are more difficult to remove; these consist of high-energy neutral species and ions with a multiple charge.

High-energy neutrals are formed by the collision of an energetic ion with a gas atom during which there is a charge exchange process. Following this process, the neutral species continues with its kinetic energy almost unaltered but without its positive charge. Neutral species cannot be focused or scanned and so they can sputter the sample in an undefined manner and disrupt the quality of the crater. The concentration of neutral species in the beam can be minimised by providing effective pumping near the source region of the ion gun. This means that the high-energy ions do not have a long path length through the high-pressure region, minimising the probability of collision with a gas atom. Some ion guns place a bend in the ion trajectory. The ions can be deflected electrostatically to follow the bend whilst the neutral species cannot and so they do not reach the sample.

An ion having a double positive charge will have twice the kinetic energy of an ion with only a single charge. The higher energy will cause greater penetration of the sample and therefore adversely affect the depth resolution. The fraction of ions with multiple charges will depend upon the conditions in the source of ions (e.g. the energy of the electrons used to ionise the gas).

#### 4.3.6.4 Information Depth

As we have seen earlier, electrons are collected from a range of depths, not just the top monolayer of material. The depth from which the electrons are collected will affect the depth resolution. In the electron spectroscopic techniques, the lower the kinetic energy of the collected electrons the smaller is the information depth and therefore the better the depth resolution. This phenomenon is illustrated in Figure 4.16 which shows spectra recorded from a germanium sample which has a thin oxide layer at the surface. The spectrum from the 3d region shows a large metallic peak whereas the spectrum from the 2p region has a much smaller contribution from the metallic peak. This is because the kinetic energy of the electrons forming the 2p spectrum is much lower and therefore a lower proportion of them are able to pass through the oxide layer.

Tilting the sample so that the detected electrons are those which leave the sample surface at more grazing angles can also control information depth. Figure 4.26 shows a comparison of Auger spectra taken at two different angles. The sample was a thin oxide layer on silicon. It is clear that the

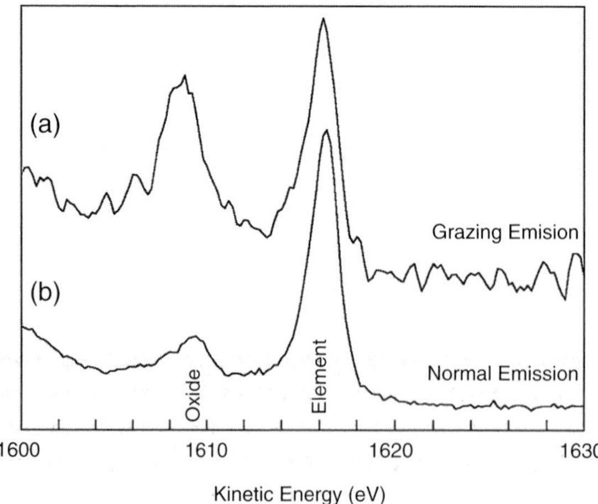


Figure 4.26 Si $KL_{2,3}L_{2,3}$ Auger spectra taken from silicon with a thin oxide layer at the surface. (a) was collected at grazing emission; (b) was collected along the surface normal. The relative intensity of the oxide peak is clearly much larger in the spectrum collected at near grazing emission.

relative contribution of the oxide to the spectrum is much larger in the spectrum taken at grazing emission angles (spectrum Figure 4.26a). This is due to the reduced information depth when electrons emitted at grazing emission angles are detected. The disadvantage of collecting at grazing emission is that the signal intensity is lower; the greater noise in the spectrum Figure 4.26a is evidence for this. The advantage of using grazing emission angles in sputter profiling is that the reduction in the information depth improves the depth resolution.

#### 4.3.6.5 Original Surface Roughness

If the surface of the sample is rough then this will affect the overall depth resolution, the roughness being maintained (or worsened) through the profile and information will therefore be collected from a range of depths at any one time.

An initially rough sample surface may become even rougher as sputtering progresses for a number of reasons:

- A rough surface may shadow parts of the sample from the ion beam
- Rough surfaces may also present many differing crystal facets to the ion beam. Each crystal facet will have a different sputter yield
- There will be a range of incidence angles between the ion beam and the sample surface. This will produce a range of sputter yields.

In some cases, it has been found that initially sputtering an inorganic sample using a beam of cluster ions can reduce the roughness of the surface without etching a significant distance into the sample.

#### 4.3.6.6 Induced Roughness

The sputtering process can cause the surface to become rough during the profiling experiment, degrading the resolution as a function of depth. This problem can be significantly reduced or eliminated by rotating the sample beneath the ion beam (azimuthal rotation) during the sputtering cycles. Azimuthal rotation is sometimes called 'Zalar rotation' after Anton Zalar who was the first to report using this method in Auger electron spectroscopy. Many commercial instruments now offer sample stages which incorporate azimuthal rotation.

#### 4.3.6.7 Preferential Sputtering

The sputter yield for each element in a multi-component sample, is likely to be different. Under these conditions, there will be roughening which may not be controllable by azimuthal rotation. Furthermore, the surface concentration of the elements will be different from the bulk concentration and so quantification of the data will be difficult, requiring the application of correction factors.

#### 4.3.6.8 Redeposition of Sputtered Material

Care must be taken to avoid the redeposition of sputtered species onto samples awaiting analysis. Furthermore, if the etch crater is small, material can be sputtered from the crater walls and redeposited within the analysis area.

### 4.3.7 Calibration

Depth profile data are presented as elemental intensity versus etch time, and a major problem in sputter depth profiling is converting this etch time scale to a depth scale. Although, in special cases, it is possible to calibrate ion sources for a particular material, it is a time-consuming procedure and, more often, the sputter rate is related to an international standard. The current standard is a $Ta_2O_5$/Ta foil with an accurately determined oxide thickness (30 or 100 nm). Thus, it is possible to report a sputter rate and an interface width for the ion gun and the conditions used in any particular piece of work.

### 4.3.8 Ion Gun Design

In addition to providing a means of compositional depth profiling in surface analysis, ion guns may be fitted to a spectrometer for other purposes. These include large area sample cleaning or as the primary beam in ion beam analysis methods such as LEISS or SIMS. The requirements for each application are slightly different and there are several different designs in common use. The guns most commonly used on electron spectrometers are those based on electron impact ionisation and those based on liquid metal ion sources (AES instruments only). Electron impact ionisation is used for monatomic argon, argon cluster, and $C_{60}$, ion guns.

#### 4.3.8.1 Electron Impact Ion Guns

Ion guns based on the electron impact source are very popular for depth profiling applications in XPS and AES. This is due to a combination of their compact design and lower cost compared with duoplasmatron designs. In addition, the ion current output of this type of ion gun at very low acceleration voltages is usually higher than that from a duoplasmatron which is important when extremely good depth resolution is required.

In this type of source, electrons from a heated filament are accelerated into a cylindrical grid where they collide with gas atoms, giving rise to the formation of ions. The ions are then extracted from the ionisation region. The kinetic energy of the ions is controlled by the magnitude of the potential applied to the grid (up to 5 kV). This ion source produces a small energy spread with only a small fraction of neutrals (i.e. unionised atoms) present in the beam. The spot sizes commercially available vary but are usually in the range 2 mm to 50 μm and depend upon both beam energy and current. The beam can usually be

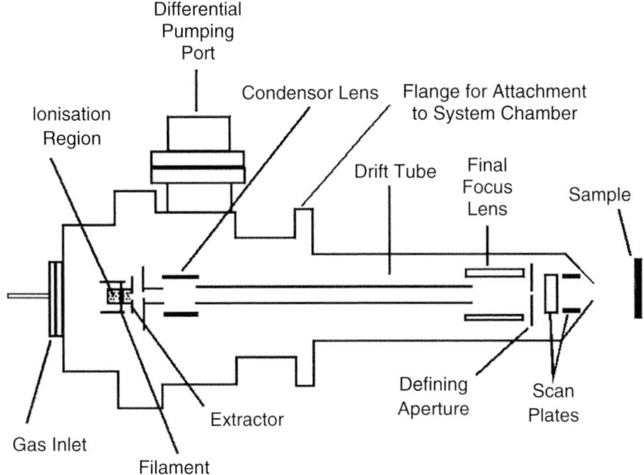


Figure 4.27 Schematic diagram of a typical ion gun with electron impact source.

scanned (rastered) over the surface to produce high-quality craters. A schematic diagram of a typical ion gun of this type is shown in Figure 4.27.

#### 4.3.8.2 Argon-Cluster Ion Guns

Argon clusters are formed by the adiabatic expansion of high-pressure (a few bar) argon gas as it passes through a carefully designed nozzle into a region of much lower pressure. The gas cools as it expands and condenses into clusters of argon atoms. To minimise the disintegration of the clusters by their interaction with other gas atoms, the volume following the nozzle is pumped rapidly to maintain as low a pressure as possible. In addition, there is a skimmer which selects clusters which are travelling in a direction parallel (or close to parallel) to the axis of the source. The skimmer also acts as a differential pumping aperture ahead of the ionisation region. These components are shown in Figure 4.28 which is a schematic diagram of the MAGCIS cluster ion gun from Thermo Fisher Scientific, reproduced here with permission.

The next section of the ion gun, the ionisation region, is very similar to that in Figure 4.27 and operates in the same way. The MAGCIS™ gun also has a gas inlet to this region which means that the gun can produce either cluster ions or monatomic argon ions.

Cluster size is selected using a magnetic/electrostatic mass filter (e.g. a Wein filter). The mass filter usually incorporates a bend so that high energy neutral species are eliminated from the beam. Cluster sizes are typically in the range of 500–2000 atoms but recent work with smaller argon clusters shows promise in certain applications.

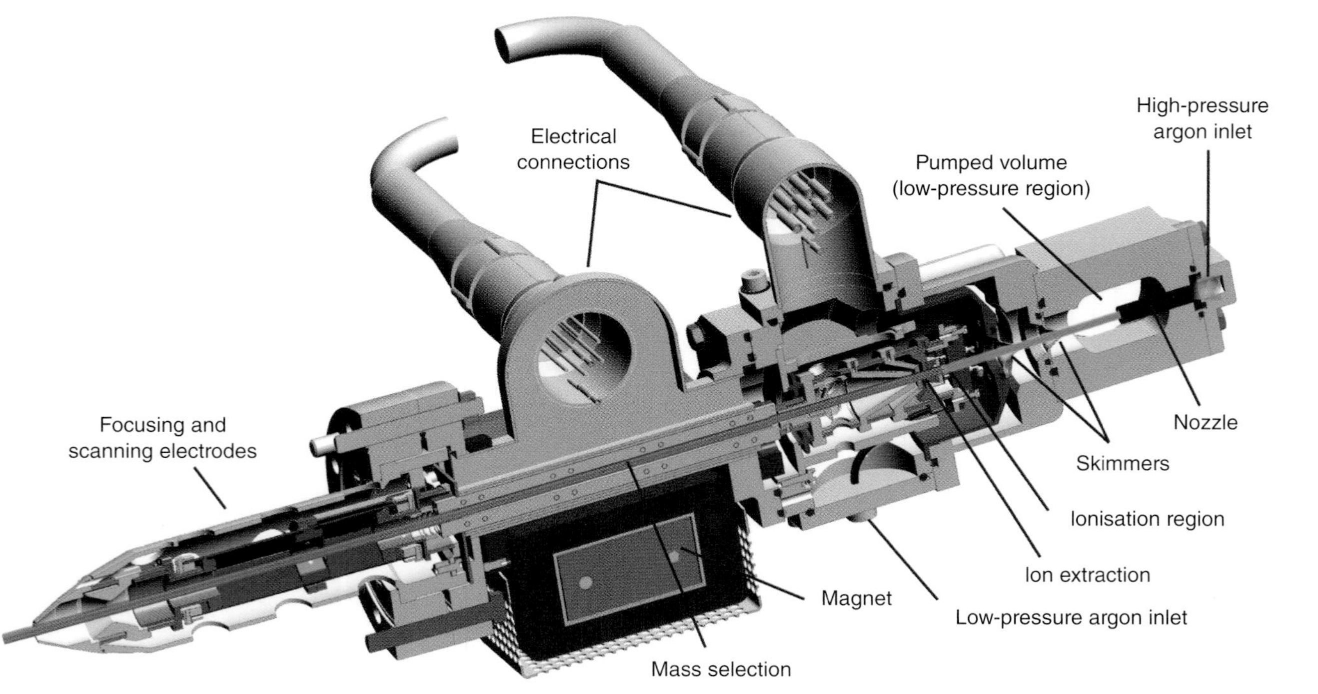


Figure 4.28 An ion gun capable of producing a beam of argon cluster ions. This design can also be used to produce a beam of monatomic argon ions.

Following the mass filter are electrodes for focusing and rastering the beam, very similar to those shown in Figure 4.27.

#### 4.3.8.3 Liquid Metal Ion Guns

For some applications, particularly when very small diameter ion beams are required, LMIGs can be used. The metal ions produced by this type of gun are usually $Ga^{+}$ but other materials have been used. These guns can produce spot sizes below 50 nm at energies above 25 keV. Although the beam current at these small spot sizes is very small (typically ~50 pA), the current densities are very high and large etch rates can be achieved over a small area. They have the added advantage that they do not impose a gas load on the vacuum system during operation. This type of ion gun is commonly encountered in SIMS analysis, occasionally used for AES profiling and rarely, if ever, used in XPS profiling. A major application of this type of gun is in micromachining, often referred to as FIB technology. Examples of its use are

1) For producing a crater in, for example, a processed silicon wafer prior to the analysis of the side wall using AES.
2) Repairing lithographic masks used in the production of semiconductor devices.

## 4.4 Sectioning

To analyse to greater depths than is practical using sputtering, it is necessary to resort to a process which removes material before beginning the analysis. Material removal can be achieved using a FIB, or by using a mechanical method two of which will be outlined here, angle lapping and ball cratering.

### 4.4.1 FIB Sectioning

As mentioned in the previous section, samples may be sectioned using a finely focused gallium ion beam (FIB). The crater produced by the rastered FIB beam has very steep, well-defined walls which can be analysed using AES to determine how the composition of the sample varies with depth.

Figure 4.29 illustrates the method for doing this if the ion gun is fitted to the analysis chamber of the Auger spectrometer. First, the sample is tilted so that the sample normal is aligned with the axis of the ion gun. The gun is then rastered over the surface of the sample to produce a crater. The size of the crater must be sufficient to allow access for the electron beam and the wall must not block the path of the Auger electrons during analysis. The crater does not necessarily have to have a completely flat bottom, it may be sloped, as shown in Figure 4.29a, or stepped. This minimises the amount of material that must be removed.

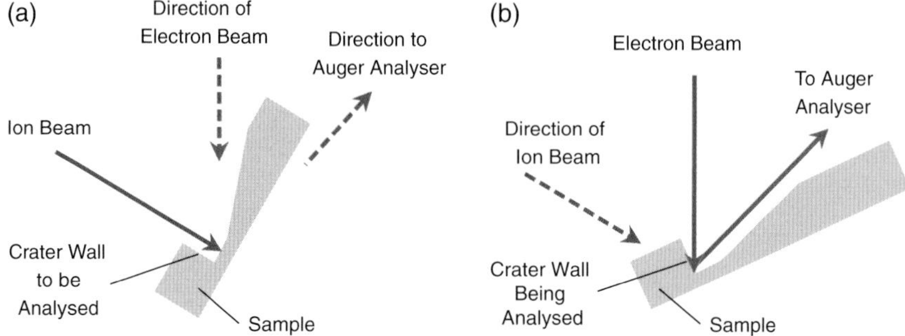


Figure 4.29 The sample alignment used for focused ion beam (FIB) sectioning and subsequent AES analysis. (a) Sample alignment to produce the crater using the FIB; (b) Alignment used during Auger analysis.

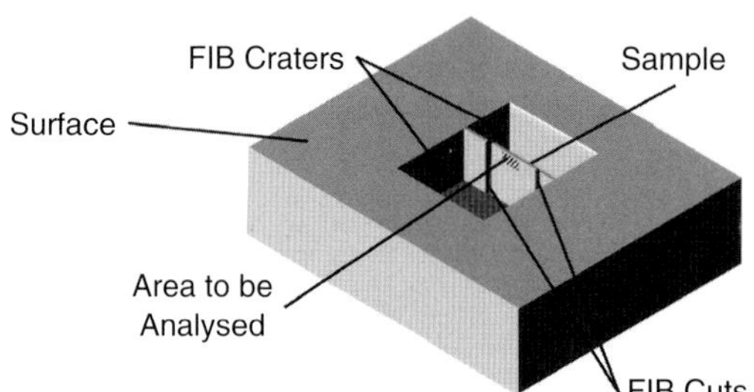


Figure 4.30 Method for producing a thin cross section for *ex-situ* Auger analysis.

Once the crater has been produced, it is often necessary to 'polish' the crater with low-energy argon ions to remove some of the damaged layer and reduce the quantity of gallium in the crater wall. The sample must then be tilted to allow the finely focused electron beam to reach the crater wall and to allow the Auger electrons to reach the analyser, Figure 4.29b.

If a stand-alone FIB is to be used the sample may be prepared using the method illustrated in Figure 4.30. Two craters are produced, separated by a thin 'wall' which is to become the sample. The wall is then cut out and placed on a conducting substrate before being loaded into the Auger spectrometer. The samples produced this way can be very thin, <100 nm in some cases. This can alleviate some of the problems associated with sample charging during Auger analysis.

### 4.4.2 Angle Lapping

When this method is employed, material is removed by polishing the specimen at a very shallow angle ($\alpha<3°$) and then introducing the specimen, with any

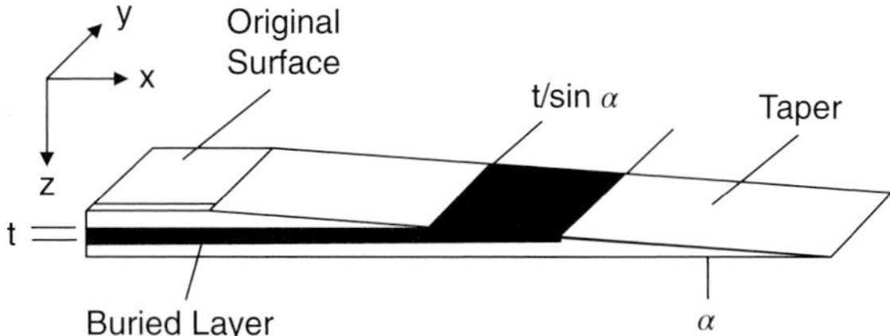


Figure 4.31 A taper section showing how the thickness of a buried layer, t, can be amplified to t/sinα.

buried interfaced now exposed, into the spectrometer. A brief ion etch to remove contamination is all that is needed prior to analysis. By carrying out Auger point analyses in a stepwise manner the variation of concentration with depth is established and it is a matter of simple geometry to convert the position of the analysis in the x-y plane to distance from the original surface, the z plane, see Figure 4.31. The main difficulty with this technique is the need to produce a very shallow taper section with a flat surface and well-defined geometry.

### 4.4.3 Ball Cratering

The problem of cutting the taper is overcome, to a large extent, by using an allied process known as ball cratering. In this process, mechanical sectioning of

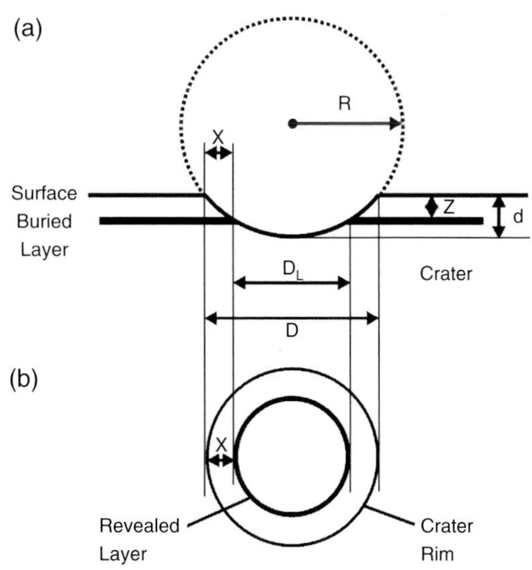


Figure 4.32 Schematic diagram of a ball cratering profile. Diagram (a) shows a cross section of a crater through a surface and a buried layer whilst (b) shows the plan view of the specimen and the revealed sub-surface layer.

the specimen is carried out by a rotating steel ball of known diameter (usually about 30 mm) coated with fine (~1 μm) diamond paste which rotates against the specimen and produces a shallow, saucer-like crater. The ball can be removed from time-to-time to assess the progress of the lapping and, on replacement, automatically 'self-centres' in the crater. Figure 4.32 shows a schematic representation of the ball cratering process.

From knowledge of the radius of the sphere, R and the crater diameter D, the crater depth d is given approximately by:

$$d = D^2/8R$$

provided that $d << R$. If x is the radial distance from the edge of the crater to the revealed buried layer then the depth of the layer, z, can be shown to be:

$$z = \frac{x}{2R}(D - x)$$

By recording Auger point analyses along the surface of the crater, a compositional depth profile can be determined.

# 5

# Multi-technique Analysis

## 5.1 Introduction

An electron gun may be added to an XPS spectrometer providing an AES capability. It is likely, however, that the lateral resolution when performing Auger measurements will be far inferior to that obtained from an instrument whose prime purpose is to make Auger measurements. Similarly, an X-ray source may be added to an Auger spectrometer/microscope giving it an XPS capability. It is unlikely that the source will have a monochromator and it is also unlikely that the spectrometer will be capable of small area XPS measurements or angle-resolved XPS (ARXPS) measurements with good angular resolution.

The prime purpose for purchasing an electron spectrometer is to analyse materials using either XPS or AES. However, many such instruments allow other valuable analytical measurements to be performed. Some of these measurements require the purchase of additional hardware (e.g. a UV lamp for UV photoelectron spectroscopy (UPS), an X-ray spectrometer for energy dispersive X-ray analysis (EDX) etc.) but others make use of the hardware that is usually present on a basic instrument. Some of these analytical techniques are described below.

## 5.2 Ultraviolet Photoelectron Spectroscopy (UPS)

The addition of a UV gas-discharge lamp allows the analyst to obtain high-intensity, valence-band spectra. Typically, these lamps are operated using helium to produce either He(I) radiation having a photon energy of 21.2 eV or He(II) radiation with energy 40.8 eV. The proportion of He(I) and He(II) photons in the beam is dependent upon the pressure of the gas in the discharge source and so can be controlled by the analyst. UV radiation derived from other noble gases may be used if required and thus a series of photon energies is available up to about 100 eV in photon energy.

*An Introduction to Surface Analysis by XPS and AES*, Second Edition.
John F. Watts and John Wolstenholme.

Synchrotron radiation is also an excellent source of high-intensity, monochromatic UV light which may be used for UPS analysis. The photon energy may be tuned to any desired value.

The information content from a UPS spectrum is similar to that available from the valence band region of an XPS spectrum. UV photons have lower energy than X-ray photons and so the kinetic energy of the emitted electrons is significantly lower. This means that UPS is even more surface specific than XPS (~2–3 nm compared with ~10 nm for XPS), making it useful for the study of adsorbed species, but surface contamination is more of a problem for UPS than for XPS. The greater intensity in a UPS spectrum allows finer detail to be observed more easily. The relative ionisation cross sections of valence band orbitals depend upon the energy of the incident photons and, in many cases, data from UPS is complementary to that from XPS valence band spectra.

UPS may be used to provide fingerprint spectra in the same way that X-ray induced valence band spectra, as shown in Figure 3.2b. It may also be used to provide information concerning the electronic structure of the valence band, as shown in Figure 5.1 which shows a spectrum from poly(dioctylfluorene) (PFO), a possible constituent of an organic light-emitting diode. More will be said about this in Section 5.4 dealing with reflection electron energy loss spectroscopy (REELS).

On an electron spectrometer which has full computer control of the gas supplies and pumps, as well as the instrumentation, it is possible to combine UPS and XPS in a depth profile experiment. This is particularly valuable if

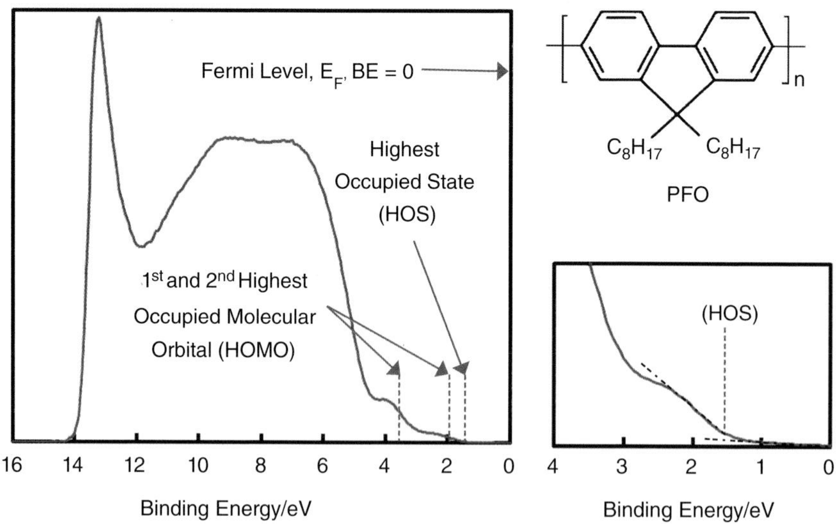


Figure 5.1 The UPS spectrum from polydioctylfluorene (PFO).

the ion source used for sputtering the sample is one that produces argon cluster ions because they minimise the damage to the molecular structure of the surface.

## 5.3 Low Energy Ion Scattering Spectroscopy (LEISS)

LEISS requires a source of low-energy, noble-gas, ions (usually <3 keV) and an electrostatic energy analyser capable of measuring the energy of the scattered positive ions. Most XPS instruments have a suitable ion source and many have an analyser capable of measuring the energy of positive species, achieved by the simple expedient of switching the polarity of the analyser.

LEISS is extremely surface specific, only detecting atoms in the top one or two atomic layers of the sample.

In LEISS, the sample is bombarded by a monoenergetic ion beam (usually $He^+$, $Ne^+$, or $Ar^+$). $He^+$ is most commonly chosen because, in LEISS, only atoms heavier than the projectile ion can be detected. The ions are scattered elastically from the atoms of the outermost layer of the solid and their energy is measured by the energy analyser. The process is illustrated in Figure 5.2 in which the noble gas ion has a mass $M_1$ and energy $E_1$ before colliding with an atom of mass $M_2$ ($M_2 > M_1$) at the surface of the sample. The collision causes the trajectory of the ion to change by an angle $\theta$ (the scattering angle) and its energy to change to $E_2$ ($E_2 < E_1$). The scattering process is elastic and so the energy lost by the ion (E1 – E2) is imparted to the sample lattice.

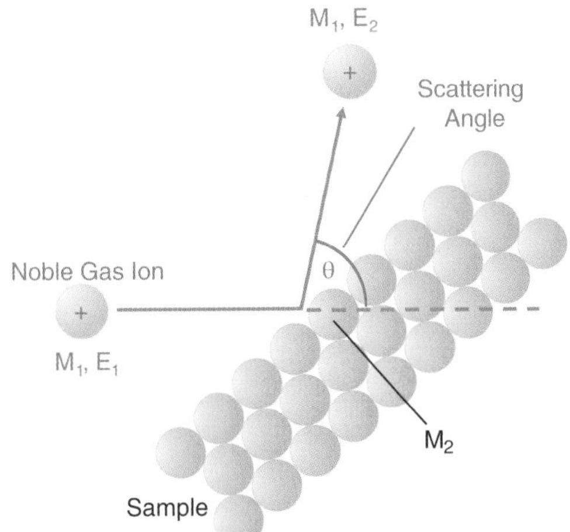


Figure 5.2 The ion scattering process.

By applying the conservation laws, it can be shown that:

$$\frac{E_2}{E_1} = \left[ \frac{\cos\theta \pm \left(q^2 - \sin^2\theta\right)^{1/2}}{1+q} \right]^2$$

where $q = M_2/M_1$.

In the special case of a scattering angle of 90°, the equation simplifies to:

$$\frac{E_2}{E_1} = \frac{M_2 - M_1}{M_2 + M_1}$$

The dependence of the energy loss of $He^+$ ions ($M_1 = 4$) and $Ar^+$ ions ($M_1 = 40$) at an energy of 1 keV ($E_1$) and scattering angle of 120° on the atomic mass of the surface atom ($M_2$) is shown in Figure 5.3. The gradient of the lines in Figure 5.3 provides an indication of the ability of the technique to separate adjacent masses. For example, using helium ions under these conditions the energy difference between the peaks $^{14}N$ and $^{16}O$ is 51 eV whilst the energy difference between $^{195}Pt$ and $^{197}Au$ is only 0.59 eV.

The ion-scattered spectrum usually takes the form of the intensity of the scattered primary beam as a function of its energy normalised to the primary beam energy ($E_2/E_1$), as shown in Figure 5.4. This shows a set of LEISS spectra

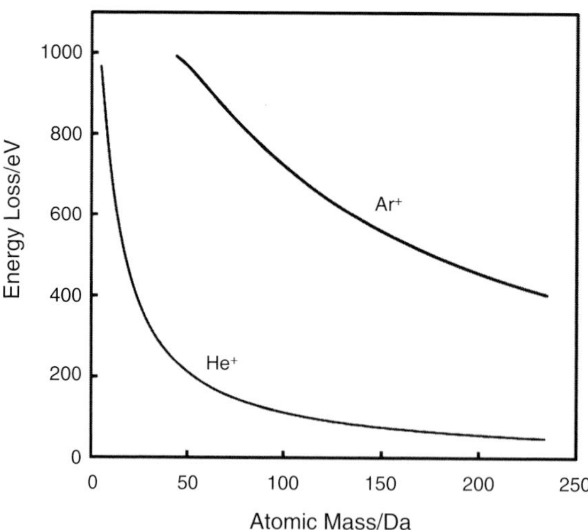


Figure 5.3 The dependence of the energy loss on the atomic mass of the surface atom. Data for both He+ ions and Ar+ ions at 1 keV with a scattering angle of 120° are shown.

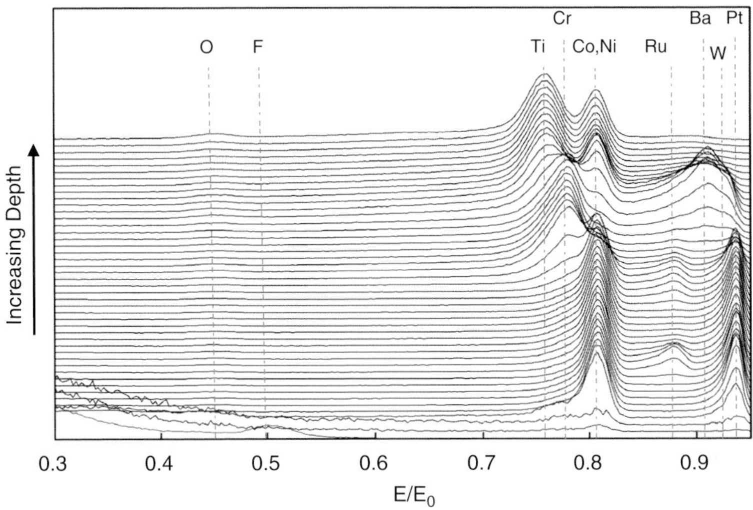


Figure 5.4 A Low Energy Ion Scattering Spectroscopy (LEISS) depth profile through the surface layers of a hard disc.

from a sputter depth profile of the layers on the surface of a computer disc. The lowest spectrum in the stack was acquired from the 'as received' sample using $He^+$ ions at an energy of 1 keV and at low current density. The gas supplying the ion gun was then switched to argon and the sample surface was sputtered using $Ar^+$ ions at a relatively high current density. The gas was then switched back to helium and the second LEISS spectrum acquired. The cycle of etching and analysing is then repeated until all of the layers have been profiled. Several distinct layers are clearly visible in the montage of spectra shown in Figure 5.4.

Measurements such as this illustrate one of the advantages of full computer control of the instrument. The ability to switch gases and control the pumping allows the operator to incorporate these tasks into the experimental procedure.

## 5.4 Reflection Electron Energy Loss Spectroscopy (REELS)

Many electron spectrometers have an electron source which, when used in combination with the energy analyser, is suitable for making some types of REELS measurement. There are two types of energy loss process, elastic scattering, and inelastic scattering, each of which can be used to complement analysis using XPS or UPS.

### 5.4.1 Elastic Scattering

When an electron strikes an atom at the surface of a solid material it imparts some of its kinetic energy to the surface atom and therefore loses that amount of kinetic energy. This assumes that the interaction does not cause an increase in the energy of any of the electrons in the solid. This process is entirely analogous to the ion scattering process described in the previous section. The same equations can be used to relate the mass of the surface atom to the energy loss suffered by the electron. In this case, $M_1$ is the mass of the electron which, of course, is much less than the mass of a hydrogen atom so hydrogen can be detected at the surface of a sample using this technique.

Figure 5.5 shows the way in which the calculated energy loss varies with the atomic mass of the surface atom. In this figure, the initial energy of the electron is taken to be 1 keV and the scattering angle is 120° (the inset shows the same data on an expanded mass scale). The difference in energy between electrons scattered from $^{1}$H and those scattered from $^{12}$C is 1.5 eV under the conditions given for Figure 5.5 whereas the energy difference between electrons scattered from $^{12}$C and those from $^{197}$Au is only 0.14 eV. This means that hydrogen can be distinguished from all other atoms, but no other pair of atoms can be distinguished.

The REELS spectra in Figure 5.6a are from three polymers, each having a different concentration of hydrogen. The spectra have been normalised to the

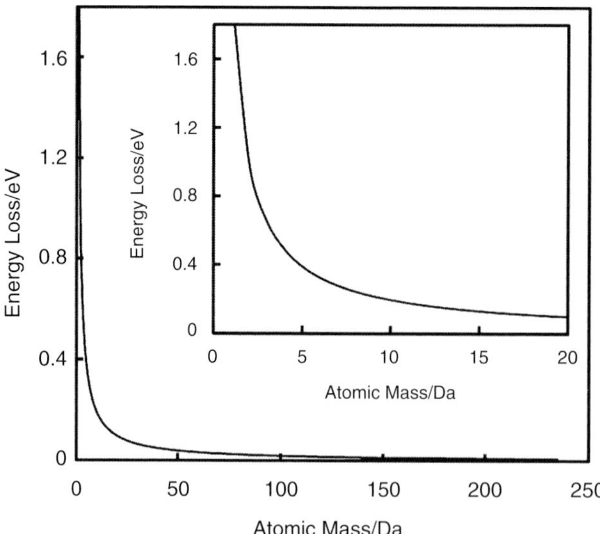


Figure 5.5 The dependence of the energy loss on the atomic mass of the surface atom. Data for electrons at 1 keV with a scattering angle of 120° are shown.

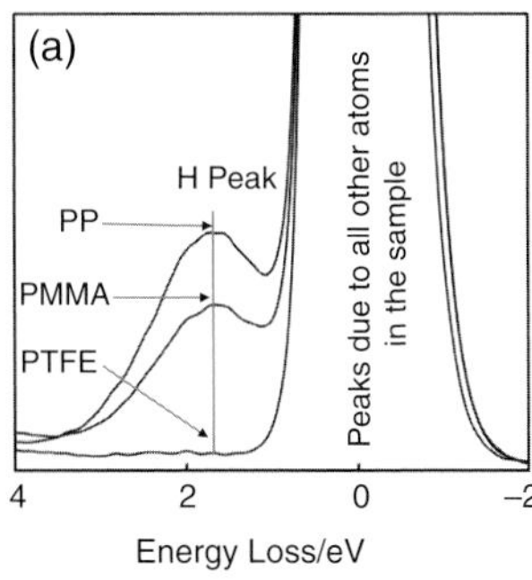


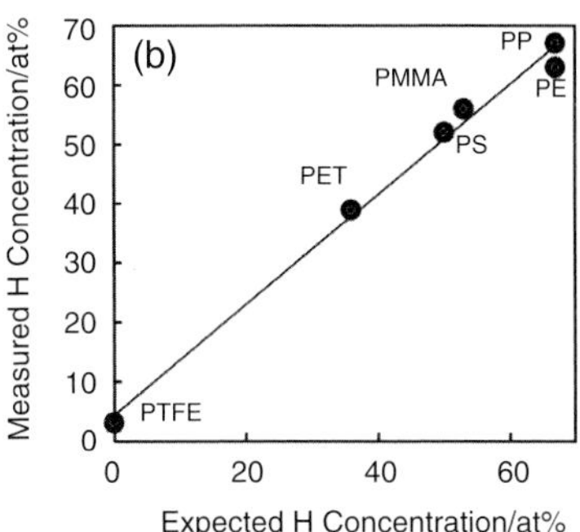


Figure 5.6 REELS spectra from polypropylene (PP), Polymethylmethacrylate (PMMA), and polytetrafluoroethylene (PTFE) showing the features resulting from elastic scattering. The peak due to hydrogen is clear on the PP and PMMA spectra.

maximum of the major peak in each spectrum. A calibration plot can be seen in Figure 5.6b. The regression line in this figure is slightly higher than would be expected, probably due to surface contaminants being richer in hydrogen than the polymers' surfaces.

The ability to detect and quantify hydrogen provides a useful complement to XPS which cannot detect hydrogen.

### 5.4.2 Inelastic Scattering

Electrons can undergo inelastic processes when they strike a solid surface. One example of an inelastic process is plasmon excitation which is a process which is confined to metals and has been mentioned in Section 3.3.8

More important analytically than plasmon excitation are losses due to the excitation of electrons in the valence band. The transitions detected are usually between bonding and antibonding orbitals, as illustrated in Figure 5.7, where the $\pi$–$\pi^*$ transitions are labelled but other features can be seen quite clearly.

A detailed knowledge of the valence band is essential, for example, in the development of materials for organic light emitting diodes (OLEDs). Polydioctylfluorene (PFO) is an example of such a material. The REELS spectrum, Figure 5.8a, can be compared with the UPS spectrum, Figure 5.1. From these spectra, it is possible to construct an energy level diagram, Figure 5.8b.

To make OLED devices, the PFO would be doped with other materials to adjust and control band structure, changing the light emitting characteristics of the device. The combination of REELS and UPS enables the doped materials to be characterised. XPS may also be used to determine the chemical composition of the material and to identify and quantify any contaminants.

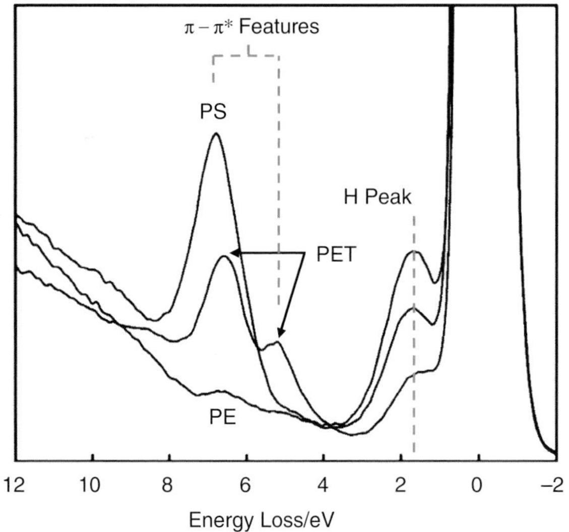


Figure 5.7 REELS spectra from polystyrene (PS), polyethylene terephthalate (PET), and polyethylene (PE). As well as showing the peaks due to elastic scattering, the peaks due to inelastic scattering are shown, these have losses greater than 4 eV.

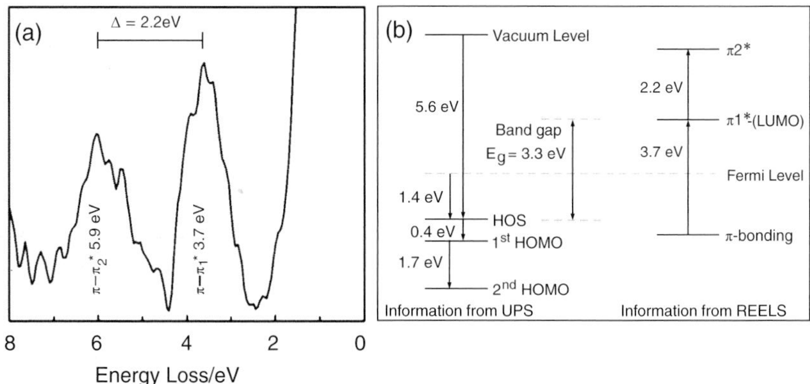


Figure 5.8 (a) The REELS spectrum from PFO; (b) the energy level diagram calculated from both the REELS spectrum and the UPS spectrum Figure 5.1.

## 5.5 Work Function Measurements

A photoelectron spectrometer may be used to determine the work function of a material. This can be achieved using a similar procedure with either UPS or XPS, although, if XPS is used, the X-ray source should be monochromated.

To make the measurement, a spectrum from the whole spectral range should be acquired so that the signal cut-off can be observed at each end of

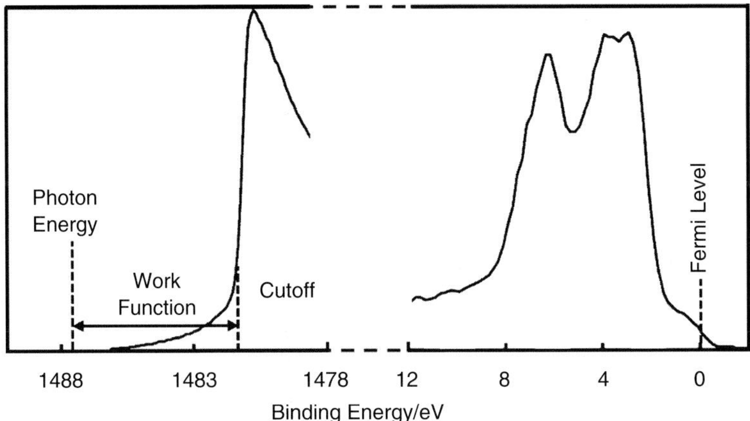


Figure 5.9 The extreme ends of the XPS spectrum from gold showing how the work function may be measured. This spectrum was acquired using a monochromated source of Al Ka X-rays (photon energy 1486.6 eV).

the spectrum. The extreme ends of an XPS spectrum of gold are shown in Figure 5.9. During the acquisition of the spectrum a negative bias of a few volts is usually applied to the sample order to accelerate the emitted electrons and to avoid complications associated with the work function of the spectrometer. The acquired spectrum is then shifted on the binding energy scale so that the Fermi level is aligned with 0 eV. The position of the high binding energy cut-off is then measured, as shown in Figure 5.9, and the difference between the cut-off energy and the photon energy is equal to the work function.

It is important that the photon energy is accurately known, and that the spectrometer's energy scale is calibrated to ensure linearity across the entire energy range. The photon energy can be measured by measuring the position of an Auger peak on the binding energy scale and adding that value to the known kinetic energy of the peak in the Auger spectrum (gold is a good reference sample for this). The binding energy scale can be calibrated from the positions of peaks acquired from standard samples (e.g. copper, silver and gold).

## 5.6 Energy Dispersive X-ray Analysis (EDX)

EDX analysis is a widely used analytical technique, most often associated with elemental analysis in the electron microscope. Unlike XPS and AES, it is the penetration of the primary beam that determines both the spatial resolution and depth resolution of the technique. An additional requirement is that the primary beam has sufficient energy to excite X-ray transitions, with a reasonable cross section, of all but the lightest elements in the periodic table. The nature of the sample will also influence the spatial and depth resolution achievable; as

Figure 5.10 An XPS instrument similar to that shown in Figure 1.13. The electron gun located at the front of the instrument allows it to be used for AES whilst the combination of the electron gun and the X-ray detector permits analysis using energy dispersive X-ray spectroscopy (EDX).

an example, if a 15 kV primary electron beam is considered a depth of analysis of ca. 1 μm will be observed in an iron sample but double this in aluminium. Similarly, the spatial resolution will be of the same order as a result of greater electron scattering in the lighter metal.

The attraction of combining EDX with AES is thus very clear. AES will provide a surface analysis (ca 5 nm) at a spatial resolution approaching that of the primary electron beam (10 nm is typical in high-end Auger microscopes). Thus, it is possible to achieve analyses in both bulk and surface modes in exact register without the need to move to another instrument or even re-position the specimen. Although EDX is perhaps most obviously considered as an adjunct to AES in a high-end scanning Auger microscope, dedicated XPS systems can function as extremely efficient Auger microscopes with the addition of a mid-range electron gun equipped with electrostatic lenses. Such a system will enable AES analysis at a spatial resolution of around 95 nm, thus the spatial resolution of the EDX data will not be significantly compromised by the use of such an electron gun. An arrangement of this type is shown in Figure 5.10 where a high-end XPS system has been fitted with a 10 kV thermally assisted field emission electron gun and a silicon drift EDX detector.

In addition to providing a conventional X-ray microanalysis in conjunction with an electron gun, the EDX detector is also able to provide an analysis by X-ray fluorescence (XRF) when operated in conjunction with a non-monochromated (twin anode) X-ray source, to make a combined XPS/XRF analysis option available. In this case, it is the bremsstrahlung component from the X-ray source that provides the energy to excite the X-ray transitions. For a typical X-ray source operating at 15 kV, it is possible to generate a perfectly acceptable X-ray spectrum up to about 12 keV, i.e. within some 2–3 keV of the source potential.

# 6

# The Sample

## 6.1 Sample Handling

Great care must be taken when preparing, transporting and mounting samples for surface chemical analysis. Surfaces are easily contaminated or damaged, by scratching, for example. Such damage can mask or even remove the intended analyte. Analysts should also be aware of the history of the sample so that the likelihood of damage or contamination can be assessed; it should be remembered that prior analysis can change the nature of the sample surface. For example, if the sample has been analysed using scanning electron microscope (SEM), the surface may have been chemically altered by the electron beam or the sample surface may have been coated with a conductive material which will prevent the surface from being analysed by electron spectroscopy.

No surface which is to be analysed by electron spectroscopy should be touched by hand. If the surface comes into contact with a plastic glove, impurities such as plasticiser or talc can be transferred to the sample and become distributed over the surface. For example, Figure 6.1 shows the XPS spectrum from a piece of clean aluminium foil in comparison with three spectra from samples of the same foil which have been touched by three different types of glove. In each case, touching the foil caused a significant increase in the percentage of contaminants, accompanied by a large decrease in the signal due to aluminium. For similar reasons, plastic bags should be avoided; aluminium foil is the preferred method of wrapping samples for transport prior to XPS or AES. Table 6.1 lists the concentrations of the elements observed in the spectra from the untouched and touched spectra and shows a significant transfer of contaminants from the glove to the surface of the foil.

Care should be taken when using tweezers to handle the sample as these can scratch the surface and so must not come into contact with the area of the surface which is to be analysed. If the sample to be analysed must be cut,

*An Introduction to Surface Analysis by XPS and AES*, Second Edition.
John F. Watts and John Wolstenholme.

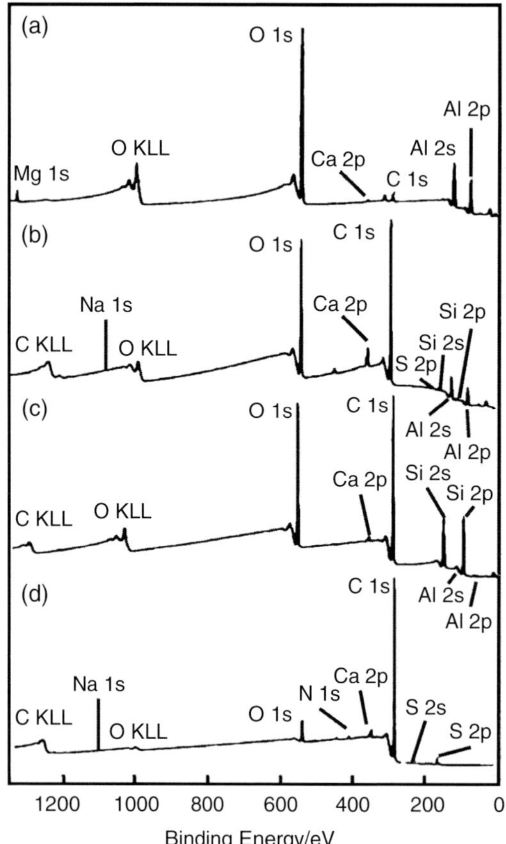


Figure 6.1 XPS spectra from (a) clean, untouched aluminium foil; (b) aluminium foil touched by a nitrile glove; (c) aluminium foil touched by one type of latex glove; (d) aluminium foil touched by a second type of latex glove.

Table 6.1 Al foil surface composition in atomic percent following glove touch-transfer testing as determined from XPS survey spectra, Figure 6.1.

| | Surface composition/atomic % | | | | | | | | |
|---|---|---|---|---|---|---|---|---|---|
| **Sample ID (Glove Tested)** | **C** | **N** | **O** | **Na** | **Mg** | **Al** | **Si** | **S** | **Ca** |
| As-received Al foil | 6.1 | ... | 52.2 | ... | 2.3 | 39.2 | ... | ... | 0.2 |
| Nitrile #1 Glove | 59.2 | ... | 23.9 | 0.1 | 0.3 | 14.2 | 0.4 | 0.2 | 1.7 |
| Nitrile #4 Glove | 91.2 | 0.4 | 5.6 | 0.1 | ... | 0.2 | ... | 1.6 | 0.9 |
| Latex #3 Glove | 55.2 | ... | 22.3 | ... | ... | 1.1 | 20.9 | ... | 0.3 |

ensure that cutting or scribing does not result in particulate contamination of the surface to be analysed. Clearly, any tool that is to touch the sample must be clean and dry. It must also be non-magnetic if there is a chance that it might magnetise the sample or sample holder.

When storing or transporting samples, they must be placed in a container which does not come into contact with the surface which is to be analysed. If the surface to be analysed is likely to react with air then all handling, transport and transfer must be done in a glove box, vacuum transfer vessel or similar.

It is also important to ensure that the samples can withstand the vacuum conditions inside the spectrometer. If volatile materials evaporate from the sample, the instrument may become contaminated. Alternatively, the pressure inside the spectrometer may rise to the point where analysis becomes impossible or not advised. Porous materials can also cause the pressure in the analysis chamber to rise.

The above is concerned with the protection of the sample surface from contamination by the environment or the analyst. However, there are materials which can pose a threat to the analyst, these need greater caution when handling. These materials include those which contain toxins, enzymes, therapeutics etc. Whenever possible, the analyst should consult the appropriate material safety data sheets. Customers requiring XPS or AES analysis from consultancy laboratories should provide COSHH[1] documentation with their samples.

A more comprehensive guide to sample handling may be found in the International Standard ISO 20579-1.

## 6.2 Sample Preparation

Many samples require analysis in the 'as received' condition and so require little or no preparation. If it is necessary to remove particulate contamination from the surface, this can be done by passing a stream of clean, dry, inert gas (e.g. nitrogen or argon) over the surface. The analyst must be sure that there are no contaminants (such as oil or particulate debris) in the gas stream. For this reason, the use of aerosol 'blow-off' sprays popular in the electron microscopy community should be avoided.

High-purity solvents (e.g. ethanol, isopropanol or acetone) may be used to remove soluble contaminants but it may be that such solvents leave a residue on the sample surface. Where possible, analar or electronic grade (>99.9%) solvents should be used. Alternatively, high-performance liquid chromatography (HPLC) grade (99.5%) may be used. Ultrasonic agitation is frequently used to assist with solvent cleaning. Wiping the sample with a solvent-soaked tissue is not recommended as impurities may be transferred to the sample from the tissue. A UV/ozone cleaner used in electron microscopy sample preparation is also useful for removing organic contamination from metallic and inorganic specimens. When dealing with biological materials, rinsing with solvents should be avoided. It is possible that the polarity of the solvent may cause a rearrangement or change in orientation of the molecules at the surface.

---

1 COSHH – Control of Substances Hazardous to Health regulations.

Ion sputtering is often used to remove a contaminant layer or to expose a buried layer. Most electron spectrometers have a suitable ion gun primarily intended for use in depth profiling measurements. When using this method for sample cleaning or preparation, the conditions (ion energy, current density and etch time) must be chosen with care to minimise the effect of the ion beam upon the surface to be analysed. The International Standard ISO 20579-2 lists the following potentially undesirable effects of sputtering as a means of cleaning or preparation:

- Atomic mixing in the uppermost layers of the sample.
- Preferential sputtering which changes the elemental composition of the surface.
- Chemical state changes.
- Changes in topography (roughening). This will affect depth resolution if a depth profile is to be measured.
- Enhanced diffusion which again causes changes to the composition of the uppermost layers.

Using a highly focused ion beam, it is possible to cut a crater whose side walls may be analysed. A liquid metal (e.g. gallium) ion source is usually used for this. This method is often followed by etching with a noble gas ion source to remove the implanted metal ions.

In recent years, argon cluster ion sources have become available. These can be an effective means for cleaning organic surfaces or removing organic contaminants from inorganic surfaces with minimum damage to the underlying material.

Other methods of revealing a new surface for analysis include the mechanical methods:

- **Fracture**. Often accomplished in the same vacuum system as the spectrometer using a specially designed fracture stage, as shown in Figure 7.1.
- **Cleaving**. Again, this may be done in vacuo.
- **Angle lapping**. This method produces a tapered section on the sample. Analysing along the length of the section produces a depth profile through the sample.
- **Low Angle Microtomy**. Similar to angle lapping but useful for polymers and biological samples; can be carried out at cryo-temperatures if sample smears.
- **Ball cratering**. This is a method for producing a deep crater, the walls of which can be analysed to produce a depth profile.
- **Scribing**. The sample may be scribed *in situ* with a sharp point to reveal a subsurface layer.

Not surprisingly, some of these methods may also be used for depth profiling, as mentioned in Section 4.4. It should be noted that the mechanical preparation techniques may give rise to particulate contamination of the sample or smearing.

## 6.3 Sample Mounting

Samples must be fixed to a suitable (usually metal) mount which can then be loaded into the spectrometer. The method used to mount the sample depends upon the nature of the sample. It is often sufficient to use a screw or simple clip to secure the sample to its holder. The holder and any screws or clips used should be metallic and nonmagnetic. A conductive material will ensure that the surface of the sample is at ground (or some well-defined) potential. Any magnetic field produced by the sample or its holder will affect electron trajectories, distorting the spectrum. Screws and clips should be positioned in such a way that they do not interfere with the incident or emitted radiation during analysis. Neither must they contact the area being analysed.

Powder samples may be pressed into a metal foil. Indium is quite soft at room temperature and is therefore a common choice. Alternatively, powders can be sprinkled onto double-sided adhesive tape, preferably tape that is conductive.

It is also possible to compress some powders into a compacted, self-supporting disc which can then be mounted with screws or clips.

A good method for mounting fibres is to lay them over a hole in the sample holder and to secure each end with a screw. The part of the sample over the hole can then be analysed without the fear of detecting signals from the sample holder.

Some XPS/AES systems will have provisions for heating or cooling the sample. The latter is particularly useful for samples that outgas to a significant amount.

## 6.4 Sample Stability

The question as to the type of samples that are amenable to XPS analysis is one that is often posed but one that cannot really be answered without a few caveats as to the instrumentation available and the type of work usually encountered within the laboratory concerned. For example, in the authors' laboratory a number of challenging samples have been successfully examined by XPS over the last five decades. These include blue cheese and other foodstuffs (potato crisps!), leaves and other botanical samples, hydrogels, tobacco, and high vapour pressure organic samples, to name but a few. The two points to be carefully considered are, whether the sample will release gaseous products during analysis to such an extent as to compromise the quality of the spectra (and indeed the safety of the spectrometer), and if there may be potential memory effects within the spectrometer. A memory effect is where a subsequent analysis of a different sample is compromised by contamination from an earlier sample. This may be one that is part of a set introduced together or indeed one processed some days before.

Fortuitously, overcoming these difficulties is fairly straightforward. In the case of samples with a high vapour pressure, the usual remedy is to chill the sample with liquid nitrogen on a cold stage. Such a facility was at one time fairly standard on most spectrometers but is now regarded as an optional extra. Such stages may be basic in nature with cooling achieved by the simple expedient of circulating liquid nitrogen through the sample holder, or more sophisticated in that a thermostatic control is used to maintain the sample at a low temperature (usually below 0 °C but well above −196 °C). This approach has worked well on many of the sample types highlighted above, but of course takes time and effort. The decision that the analyst must make is whether such effort is justified on the samples to hand. With some samples that possess a high surface area, such as wood, it may be necessary to pump in an auxiliary chamber, such as a fast entry air lock, for an extended period of time. Common practice is to load samples the day before analysis, and is undertaken to allow for such extra pump down time. The feasibility of such approaches depends on local operating practices, but in many years of carrying out XPS analysis in the University of Surrey laboratory only one sample has totally defeated efforts of analysis; this was a solid organic pigment which sublimed on gentle pumping!

The alternative approach is the use of near ambient pressure XPS (NAPXPS) (see Section 2.6) where pressures in the region of several mbar can be accommodated in the environs of the sample. This, of course, requires ready access to a NAPXPS system. Examples of the use of this approach in samples that are less vacuum compatible than one would like are provided in Chapter 7.

The issue of memory effect in XPS is less of a problem once it has been identified. In some cases, the system will 'pump itself clean' over a few days of inactivity (such as a weekend), but the more rigorous way to ensure system cleanliness is to carry out a short system bake. In modern compact systems this is readily achieved overnight as there is very little dismantling of external components required. Although a high-end instrument will require more dismantling and subsequent re-assembly it should not take more than one day (including an overnight bake). The main challenge as far as memory effects are concerned is identifying them in the first place. If suspected, a useful diagnostic is to etch a metal standard clean (oxide free) in the spectrometer and assess any build-up of extraneous matter over the period of a few hours.

Whilst XPS is relatively tolerant of sample stability, AES is very sensitive to issues with a sample that will influence stability under the electron beam. Even a small amount of sample charging may be sufficient to influence the energy distribution of the emitted electron spectrum to the point where it contains little, if any, useful information. Localised charging of components within the chamber of an Auger electron microscope may have a similar effect. This may arise, for example, from the deposition of material from sputter depth profiles, after extended use. The solution here is relatively straightforward, a silver sample sputtered for an hour or so will deposit a conducting film on such layers and all is well. In summary, for AES it is essential that samples are vacuum compatible and do not outgas or degrade on electron irradiation. Insulating samples are challenging but the use of a low-energy argon ion beam for charge neutralisation or a low angle of incidence of the primary beam (as noted in Section 6.6.2). may control the charging and allow a satisfactory spectrum to be recorded.

## 6.5 Contamination and Damage During Analysis

There are many potential causes of damage and contamination to the sample surface and the analyst must be aware of these and their affect upon the electron spectrum. Some causes of contamination are listed below:

- Contact with tools.
- Contact with gloves.
- Exposure to gases (including the analyst's breath).
- Exposure to the instrument's vacuum. Water vapour, plasticiser etc. desorbed from previously analysed samples can become readsorbed (the memory effect).
- Exposure to magnetic materials. This is not strictly a form of contamination but if the sample becomes magnetic then the electron spectrum will be affected.

The possibility of sample contamination during storage must also be considered.

During analysis, samples are exposed to ultra-high vacuum and high-energy particles and/or photons. All of these are capable of inducing undesirable changes to the sample (damage) and cause changes in the electron spectrum:

- Desorption of gases into the vacuum has been mentioned above.
- Incident X-rays and electrons can cause changes to the bonding or oxidation state within the sample. Such processes as degradation, polymerisation, reduction may occur.
- An X-ray flux can cause the sample temperature to rise which may in itself be sufficient to alter the nature of the sample.
- In AES analysis, the potential for damage to the analysed area due to these causes is greater than for XPS because the electron beam used for AES has a much higher flux density than the X-ray beam used in XPS.
- Typically, a non-monochromated X-ray source is brought very close to the sample during XPS analysis. This can result in damage from two causes; radiant heat from the source can cause sample heating and high-energy electrons emitted from the X-ray window can cause damage in similar ways to those caused by X-rays.

If it is thought that a sample is susceptible to damage during analysis then parts of the analysis may be repeated to check that there is no change. To minimise damage during analysis, minimise both the flux density of the input radiation and the time during which the sample is exposed to the beam consistent with getting a satisfactory signal to noise ratio in the data.

## 6.6 Controlling Sample Charging

XPS relies upon the emission of electrons from the surface of the sample. If this sample is an insulator, then the surface will develop a positive charge. Such a charge will decrease the kinetic energy of the emitted electrons and shift the spectrum along the binding energy scale. If the charging is non-uniform (as is highly likely) then the peaks in the spectrum will broaden and the electrostatic field near the surface of the sample will affect the trajectories of the emitted electrons, often to the point where no photoelectrons reach the analyser. It is therefore essential that sample charging is carefully controlled. It is not essential that the sample surface is maintained at ground potential so long as the potential remains constant and uniform during the analysis.

As will be seen, the situation is more complex for AES because electrons form both the incident and the emitted radiation.

### 6.6.1 Sample Charging in XPS

When photoemission is excited by a non-monochromatic X-ray source, there is usually a sufficient number of low-energy electrons available in the neighbourhood of the sample to neutralise the sample effectively and allow high-quality XPS spectra to be obtained.

When monochromatic X-ray sources are used, these low-energy electrons are not produced in such large numbers near the sample and so neutralisation does not take place. Indeed, because the X-ray line width from a monochromator is much narrower, the need for effective charge compensation is even greater.

When charge compensation is necessary, it is normal practice to flood the sample with low-energy electrons. It is not usual to attempt to balance the charge exactly, an excess of electrons is used to produce a uniform negative potential of known magnitude at the surface of the sample. The peaks can then be shifted to their correct positions during data processing. This technique minimises the risk of differential or non-uniform charging.

The electron beam used for charge compensation should be of low energy to minimise the risk of damage to the surface of the sample but must be of a sufficiently high flux to adequately compensate for the charging. Typically, the energy of the electrons is less than 5 eV.

When using a focusing X-ray monochromator there is an additional complication. The area of the sample being struck by the X-rays will charge in the

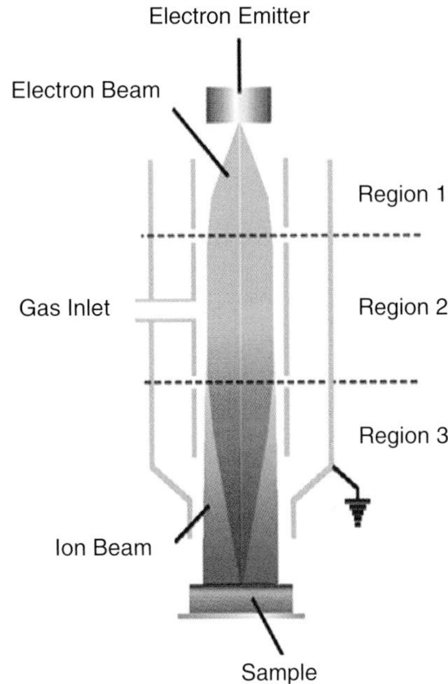


Figure 6.2 Schematic diagram of the combination ion/electron gun used for charge compensation.

positive direction but it is not uncommon for the surrounding area to have a negative charge. This negative charge can prevent the low-energy electrons, intended for charge compensation, from reaching the area being irradiated by the X-rays. One method for getting around this problem is to use a combined beam of low-energy electrons and low-energy ions.

At least one manufacturer offers a combination ion/electron gun, shown schematically in Figure 6.2. This design uses a source of low-energy electrons, with a small energy spread (~0.3 eV). The electrons are focused to a spot on the sample and aligned with the X-ray beam. The electrons are accelerated into the middle element of a three-element electrostatic lens (region 2 in the diagram). Argon gas is introduced into this region of the gun where some of the argon atoms are ionised by the high-energy electrons. The remaining electrons are then retarded as they enter region 3 of the lens and are focused on the sample. The ions produced in region 2 are accelerated into region 3 and then strike the sample as a diffuse beam.

This method for controlling sample charging has proved to be successful for a very wide range of sample types.

For some samples, it is sufficient simply to place a grounded metal grid or disc with an aperture in the centre over the area to be analysed. The analysis can then be accomplished by analysing through the grid or aperture. Wrapping the sample in conductive foil but leaving a hole over the area to be analysed achieves a similar effect. Conductive paint may also be used. Painting a stripe between the sample holder and the surface of the sample is sufficient in some cases.

### 6.6.2 Sample Charging in AES

For Auger analysis, however, the use of an electron beam dictates that for routine analysis the sample should be conducting and effectively earthed in addition to the vacuum compatibility requirements outlined above. As a guide, if a sample can be imaged (in an uncoated condition) in an SEM without any charging problems, a sample of similar type can be analysed by Auger electron spectroscopy.

Some of the methods described in the previous section (grids, apertures, wrapping in foil etc.) may also be employed for AES. In many cases, the use of low-energy ions can also be effective.

The analysis of insulators such as ceramics by AES is quite feasible but its success relies heavily on the skill and experience of the instrument operator. Such analysis is achieved by ensuring that the incoming beam current is balanced exactly by the combined current of emitted electrons (all secondaries including Auger electrons, backscattered and elastically scattered electrons etc.). Figure 6.3 shows how the secondary electron yield ($\delta$) varies with incident beam energy. The secondary electron yield is defined as the ratio of the total number of electrons emitted from a sample to the number of electrons incident at a given energy and angle. It can be seen that the curves for each angle pass through 1 at two points as the energy is increased. For the curve representing $\delta$ at an angle 0° these points are labelled $E_1(0°)$ and $E_2(0°)$. At these points, the total electron current arriving at the sample is equal to the total electron current leaving the sample and so, at this energy, there is charge balance and it should be possible to obtain a good quality Auger analysis.

In this example, $E_2(0°)$ is approximately 2.8 keV. At this energy, the spot size of the electron beam is far greater than the minimum for a typical electron gun and so the lateral resolution of the analysis is severely impaired. This can be overcome to some extent by tilting the sample. In this example, $E_2(50°)$, is somewhere in the region of 5.5 keV. At this energy the spot size capable of being produced by the electron gun is smaller than that at 2.8 keV but is likely to be well above the smallest that the gun can produce. In addition, the analysis area will be enlarged in one direction by virtue of the sample tilt.

As can be seen from Figure 6.3, the value of $\delta$ is less than 1 at high energy. This means that the sample will charge in the negative direction and so a beam of low-energy ions can be used to achieve a stable and uniform surface potential.

If the insulating material being analysed is thin (a few hundred nanometres) and is in contact with a conducting substrate then charging can be much less

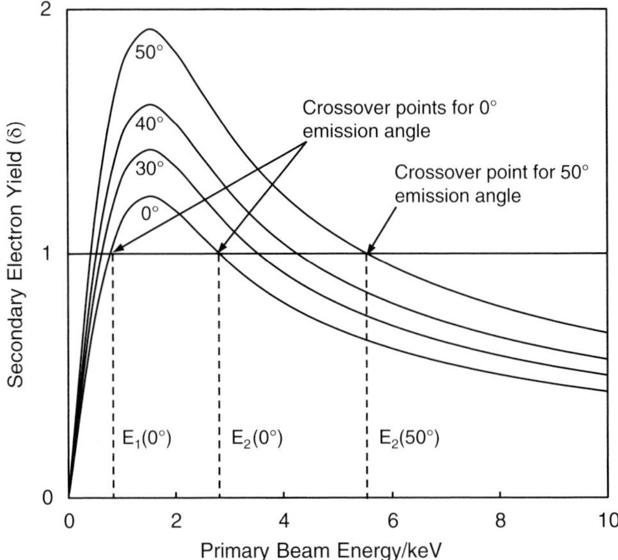


Figure 6.3 Secondary electron yield as a function of beam energy at four sample angles for a typical insulating sample.

severe. If the thickness of the layer is small compared with the penetration depth of the electron beam, then most of the beam current will pass into the conducting substrate. Focused ion beam (FIB) sectioning (see Section 4.4.1) can be used to produce a very thin cross-sectional sample which can be placed on a graphite substrate before being analysed by AES. Very good, high-resolution SAM images may be produced using this method (see Section 7.6.1.).

# 7

# Applications of Electron Spectroscopy in Materials Science

## 7.1 Introduction

So far, in this text, we have been concerned with the practice of electron spectroscopy and the interpretation of the resultant spectrum. We will now consider the way in which it is possible to make use of these surface analysis techniques to provide information which furthers our knowledge in a particular discipline. Although X-ray photoelectron spectroscopy (XPS) and Auger electron spectroscopy (AES) together with scanning Auger microscopy (SAM) are used widely in all branches of pure and applied sciences – as well as for trouble shooting and quality assurance purposes – the only area that we will consider in this chapter, is their use in materials science investigations. If we subdivide this group, it is possible to identify the following applications headings; metallurgy (including surface engineering), corrosion, ceramics, microelectronic and semiconductor materials, polymers, adhesion, nanotechnology, biology, and energy. We shall consider each of these areas in turn, representative references for each are listed in the bibliography which will provide interested readers with further examples and guidance in their field of interest.

## 7.2 Metallurgy

In the field of metallurgy, it is Auger electron spectroscopy that has proved to be the most popular technique, and with good reason. The majority of investigations are concerned with the diffusion of elements within metallic matrices. This may take the form of interdiffusion of metallic coatings with the substrate or of the surface segregation of minor alloying elements on heating in oxidising or reducing atmospheres. However, the major contribution of Auger electron spectroscopy to metallurgy, especially in the early days of the development of surface analysis, was the investigation of grain boundary segregation and

*An Introduction to Surface Analysis by XPS and AES*, Second Edition.
John F. Watts and John Wolstenholme.

embrittlement in structural steels. In addition, both AES and XPS have been used in 'quality assurance' and sometimes 'forensic' roles to ensure (for example) that rolled steel sheet is of adequate cleanliness or to identify surface phases which lead to poor compaction in powder metallurgy processing.

### 7.2.1 Grain Boundary Segregation

The embrittlement of structural steels result from the aggregation of certain elements, present in very low or trace quantities in the bulk material, at the prior austenite grain boundaries. The grain boundaries are weakened to such an extent that they become the preferred fracture path, with catastrophic effects on the mechanical integrity of the material. The elements most widely investigated are phosphorus and sulphur, but the effect is brought about by, and has been studied for, silicon, germanium, arsenic, selenium, tin, antimony, tellurium, and bismuth. The quantity of grain boundary segregant involved is necessarily very small, probably sub-monolayer, and located at the grain-boundary within a material of grain-size of about 100 μm or less. Thus, the need for surface specificity and reasonable spatial resolution is immediately apparent. In order to measure the quantity of segregant at the interface, the steel must be fractured in an intergranular manner, usually at, or near, liquid nitrogen temperature. This must be carried out within the ultra-high vacuum (UHV) environment of the spectrometer to prevent oxidation of the iron matrix and subsequent obliteration of the small signal from the segregant. Nowadays, most manufacturers offer such a fracture stage for their Auger microscopes, the more sophisticated having the ability to analyse both fracture surfaces; an example is shown in Figure 7.1. All rely on fracture by a fast, three-point bend configuration (similar to the geometry referred to by the metallurgists as an Izod Test). Scientists requiring controlled strain-rate fracture must still resort to building their own devices.

The surface morphology generated by such low temperature fracture is sometimes a mixture of regions of intergranular and transgranular failure. This provides a convenient comparison between the matrix composition (transgranular) and a grain boundary analysis from the region of brittle failure. However, most metals, particularly steels, will provide a brittle failure mode when fractured at cryogenic temperatures approaching 80 K.

The AES spectra of Figure 7.2 show the effects of heating a Cr2.25Mo1 steel with a level of sulphur and phosphorus of 0.5 wt% at temperatures of 480–560 °C for times of 20–50 days. The three spectra shown represent the times to establish equilibrium concentrations of phosphorus at these temperatures, respectively 18, 22, and 27 at.% for temperatures of 560, 520 and 480 °C, respectively.

The presence of a very weak O *KLL* feature at ca. 510 eV convolved with the Cr *LMM* triplet indicates that even under clean UHV conditions some oxidation of

Figure 7.1 Fracture stage attached to a modern Auger electron spectrometer.

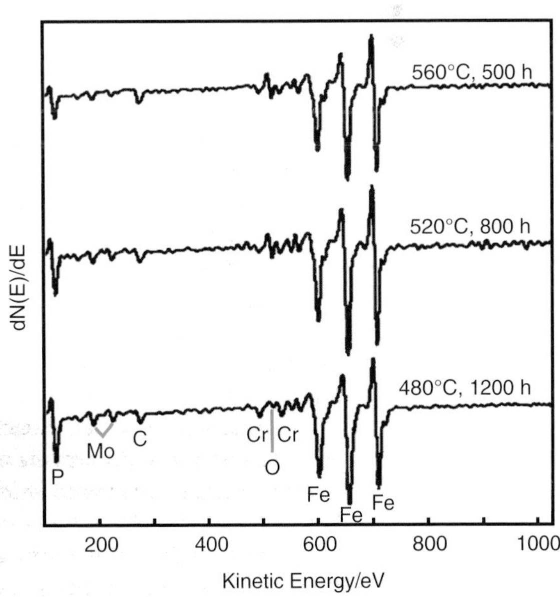


Figure 7.2 Differential Auger spectra from regions of a 2.25Cr1Mo steel fractured at liquid nitrogen temperatures, following heat treatment for the temperatures and times indicated. (*Source:* reproduced from Wu et al. [2008],with permission).

the surface (invariably the chromium) may occur. If the sample had been fractured in air, the oxide formed would come close to obliterating the phosphorus segregant at the surface.

Studies have been made of many systems that exhibit grain boundary segregation and the underlying theory is now well developed, mainly as a result of systematic studies undertaken at the National Physical Laboratory (NPL) in the UK. Thus, the extent of grain boundary segregation may be predicted by the following equation for dilute levels of segregant in the matrix:

$$\beta = \frac{K}{X_C^0}$$

Where β is the grain boundary enrichment ratio, $X_C^0$ is the solid solubility of segregant in the matrix and K = exp(−ΔG/RT), ΔG is the free energy of segregation. This equation describes a large number of experiments undertaken on many systems all indicating that the degree of enrichment is dependent on solid solubility over a very wide range of concentration (from 100 ppm to 100%), see Figure 7.3.

Thus, in the field of segregation and embrittlement, AES has not only provided a technique that enables the level of segregant to be qualitatively assessed, it has, as a result of such measurements, enabled the development of an underlying theory which predicts such a phenomenon very accurately.

### 7.2.2 Electronic Structure of Metallic Alloys

Although it is Auger electron spectroscopy that has been widely used in metallurgical studies because of its superior spatial resolution, XPS is used where information at the chemical or electronic level is required. Alloy design or development has traditionally been carried out based on both empirical or systematic methods and experimental and calculated phase equilibria, as presented in the equilibrium phase diagram. Thermodynamic modelling has been applied quite successfully over the last two decades but is essentially phenomenological in nature and cannot provide adequate information on the subsequent electronic changes (such as charge redistribution) that occur upon alloying. These data are accessible by making so-called first principle calculations but are not easily able to provide phase diagrams of sufficient accuracy. Furthermore, there is a serious need for experimental evidence regarding the phenomena accompanying alloy formation such as charge transfer and redistribution. This can be achieved using the Auger parameter of the solvent and solute elements of the alloy and the linear potential core model, developed by Thomas and Weightman, to relate the Auger parameter to changes in electronic structure. The description of the model is outside the scope of the text, and the reader is referred to papers cited in the bibliography, but examples from a Ti-Al-V alloy is given below to indicate the power of the technique.

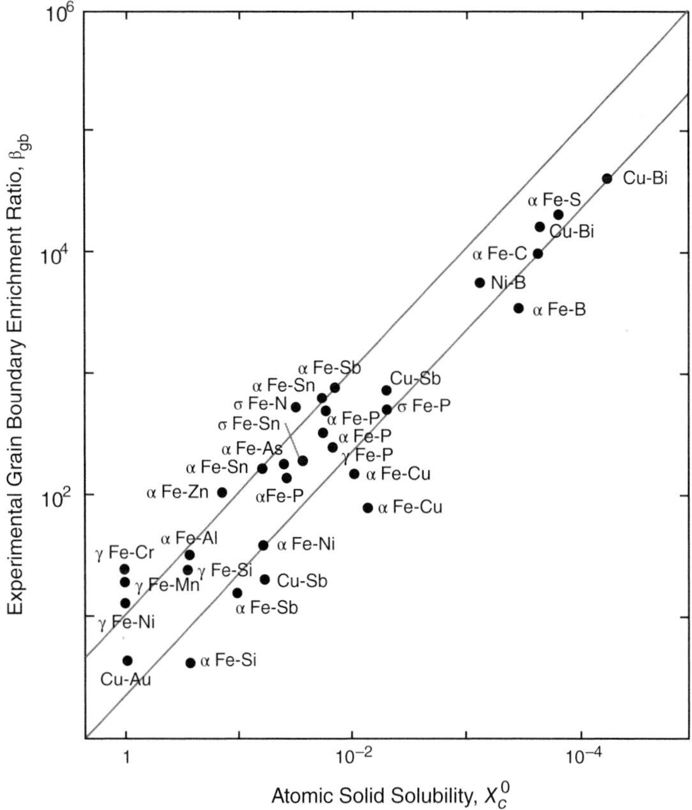


Figure 7.3 Grain boundary enrichment (β) versus inverse solid solubility for a range of binary alloys. (*Source:* reproduced from Seah [1980], with permission).

In order to obtain reliable information from Auger parameter data one must use core-like Auger transitions, i.e. ones where all three electrons involved in the Auger process originate in core-like orbitals rather than the degenerate band structure of the atom. In order to achieve this with metals heavier than magnesium it is necessary to resort to a high energy X-ray source, such as the Cr Kβ source, which fulfils fourth order reflections in a conventional Al Kα monochromator. Such an arrangement provides a photon energy of 5946.7 eV and a line width of approximately 1.6 eV, and, although rather weak, is convenient and available in a conventional Al Kα monochromator by the simple expedient of replacing the aluminium anode with a chromium one. It should be noted that very recently a system equipped with a Cr Kα source (h$\nu$ = 5417 eV) has become commercially available. The investigation of the TiAlV alloys, made use of a full set of spectroscopic data from Al, Ti, and V transitions (1s, 2p, and *KLL*) relative to the pure metal. Final and initial state Auger parameters (α and ζ – see Section 3.3.4) were calculated using the 1s-2p-*KLL* energies, for a range of alloys, and are presented along with the primary data, in Table 7.1.

It has been suggested that the Auger parameter changes are a measure of the screening efficiency of a system in response to the presence of a localised core hole. Considering that metals are characterised by perfect screening, reduced screening means that the atoms experience a 'less metallic' environment. The Auger parameter shifts (both initial and final state) of the elements of interest between pure metal and alloy in Table 7.1 suggest that the metal with sp valence configuration (i.e. Al) is better screened in the elemental solid than in both the binary and ternary alloys. The difference in the Auger parameter values of Al between the pure element and in the alloys is bigger for the alloy systems with ordering tendency (Ti-Al, Ti-Al-V) than a system with no ordering tendency such as V-Al. Furthermore, the Al Auger parameter is lower for all alloys compared with pure Al, indicating a lower intra-atomic and extra-atomic relaxation or a reduced screening efficiency. The use of such a high-energy source is a relative luxury but similar qualitative, rather than quantitative, observations can be made with a conventional source. Figure 7.4 shows the Al 2p spectra for a series of Al-Ti alloys compared with pure Al. The

Table 7.1 Differences in core level binding energies of Al, V, and Ti, and initial and final state Auger parameters, for the alloy relative to the pure metal. The negative sign indicates a decrease in the value for the alloy compared with the pure metal.

| | Ti-10Al | Ti-20Al | Ti-30Al | Ti-50V | Ti-25V-5Al | Ti-20V-40Al |
|---|---|---|---|---|---|---|
| **Aluminium** | | | | | | |
| **Δα** | −0.9 | −0.7 | −0.7 | | −0.7 | −0.6 |
| **$\Delta E_B$ Al1s** | −1.1 | −1.0 | −0.9 | | −1.0 | −0.8 |
| **Δζ** | −2.1 | −1.8 | −2.1 | | −2.5 | −2.0 |
| **$\Delta E_B$ Al2p** | −0.6 | −0.6 | −0.7 | | | |
| **Titanium** | | | | | | |
| **Δα** | −0.1 | 0.0 | +0.1 | −0.3 | −0.2 | −0.1 |
| **$\Delta E_B$ Ti1s** | −0.3 | −0.1 | −0.1 | −0.3 | −0.2 | −0.1 |
| **Δζ** | −0.4 | −0.2 | −0.5 | −0.5 | −0.3 | −0.1 |
| **Vanadium** | | | | | | |
| **Δα** | | | | +0.3 | 0.0 | −0.1 |
| **$\Delta E_B$ Ti1s** | | | | +0.2 | −0.1 | −0.1 |
| **Δζ** | | | | +0.3 | −0.2 | −0.5 |
| **Structure** | A3 (hcp) | α2-DO19 | α2-DO19 | A2 (bcc) | | B2 (bcc) |

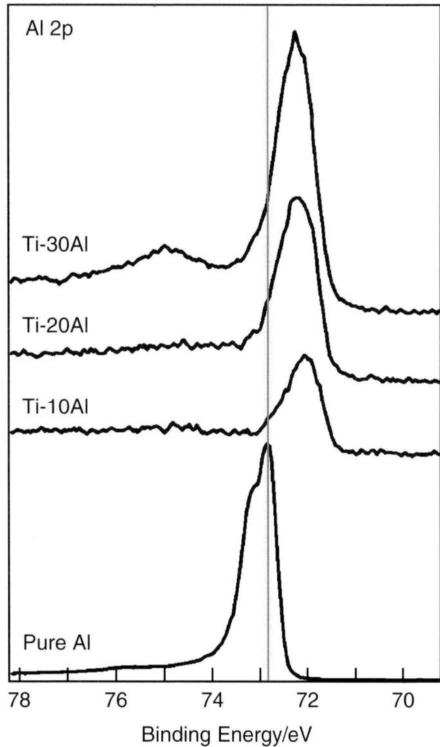


Figure 7.4 Al 2p spectra for a series of Al-Ti alloys compared with pure Al. (*Source:* reproduced from Diplas et al. [2001], with permission).

chemical shift experienced by the Al in the presence of Ti, compared to the binding energy of the pure Al 2p electrons is clear.

### 7.2.3 Surface Engineering

The modification of a metal surface to provide specific corrosion of tribological properties has been carried out for more than a century but is nowadays recognised as a discipline in its own right: surface engineering. Both AES and XPS are widely used in the analysis of metallic and non-metallic coatings and the analysis complexity varies from the straightforward to the extremely difficult. The former is illustrated by the analysis of an electrodeposited zinc coating on a steel substrate. The coating was known to be several tens of micrometres thick and the sample was prepared by ball cratering (see Section 4.4.3). Following introduction into the spectrometer the surface was briefly sputtered to remove adventitious contamination and a linescan recorded from the crater edge to the exposed steel substrate. Using a simple geometrical manipulation, the lateral distance can then be converted to depth and plotted as the depth profile shown in Figure 7.5. The chromium present at the outer surface is the

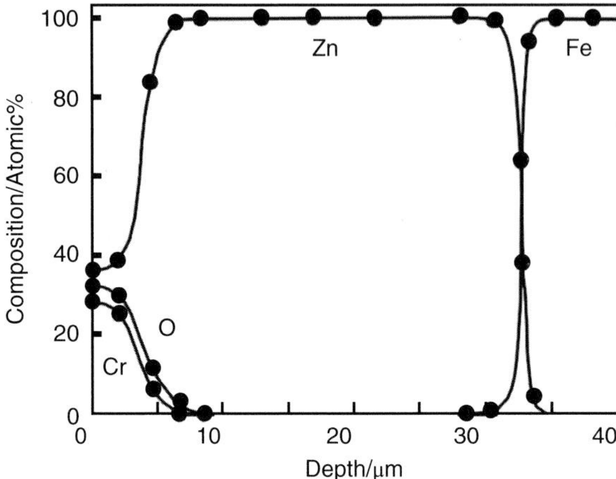


Figure 7.5 Depth profile through a zinc coating on a steel substrate, from a ball cratering experiment.

result of the use of a passivation (chromate) treatment following zinc deposition. The zinc deposit is seen to be about 30 μm thick.

Titanium nitride (TiN) and titanium boronitride (Ti-B-N) coatings are very attractive propositions as wear resistant coatings. As with many such investigations it is often desirable to correlate the coating composition with performance in some empirical test to establish the optimum coating composition. In the analysis of TiN coatings by AES there is a potential problem as a result of the overlap, at a kinetic energy of 309 eV, of the N$KL_{23}L_{23}$ with the Ti$L_3M_{23}M_{23}$ that is usually studied, as shown in differential Auger spectra of TiN, Ti, and $TiB_2$ (Figure 7.6).

To overcome this problem, it is necessary to resort to using the Ti$L_3M_{23}M_{45}$ at a slightly higher kinetic energy (c. 420 eV) and establishing a calibration curve using samples of known composition (determined by XPS), which can be plotted relative to the ratio of peak to peak intensities of (N$KL_{23}L_{23}$ + Ti$L_3M_{23}M_{23}$)/(Ti$L_3M_{23}M_{45}$) as shown in Figure 7.7.

In this case, it is helpful to degrade the analyser resolution to negate the effect of fine structure in the Auger spectrum. An alternative approach, which relies on the presence of chemical effects in Auger spectra recorded at high spectral resolution, can be achieved by simply using the Ti$L_3M_{23}M_{45}$ transition alone. This region of the spectrum, recorded in the direct rather than differential form, is shown in Figure 7.8 for a series of $TiN_x$ coatings, and the evolution of a Ti$L_3M_{23}$ *hybrid* feature at a slightly lower kinetic energy is well resolved in the spectrum.

This hybrid feature is a result of the drawing the Ti 3d valence electrons (close to the Fermi level) towards the N 2p electron situated below the Fermi level and the formation of a hybrid bond. The involvement of the feature in

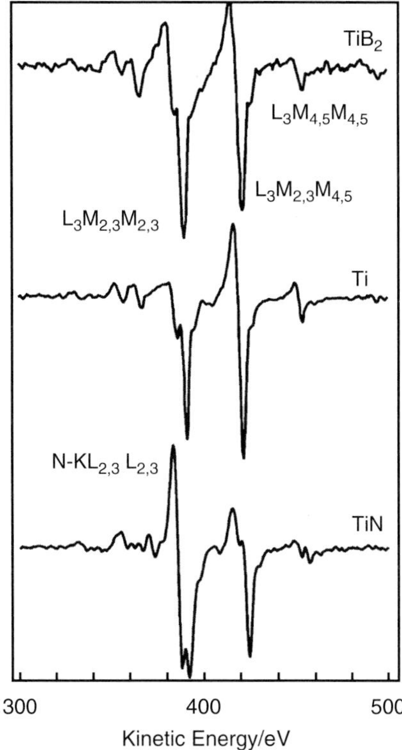


Figure 7.6 Auger spectra of Ti, TiN, and $TiB_2$. (*Source:* reproduced from Baker [1995], with permission).

the Auger transition is then seen as a distinct feature in the Auger spectrum, which increases in intensity with increasing nitrogen content. In a similar manner to the previous example, it is possible to plot a calibration curve of peak area ratio Ti$L_3M_{23}$*hybrid*/Ti$L_3M_{23}M_{45}$ which is arguably more accurate than the method described using peak to peak heights of the differential spectra, above. The slight drawback is the need to carry out careful peak fitting and background removal to obtain an accurate estimation of the ratio of components.

In the case of the boronitride, it is possible to unravel the complexity of the TiN, $TiB_2$, and BN phases in a complex coating by study of the B1s XPS spectra as shown in Figure 7.9.

The relative proportions of the three phases are in very good agreement with those predicted by the phase diagram even though the coating in question was deposited under conditions very far from equilibrium! The case for titanium borocarbide coatings is, however, very different. In this coating the phase diagram indicates a three-phase structure of $TiB_2$, TiC, and carbon, the carbon is thought to be in the form of diamond-like carbon (DLC). The C 1s spectrum

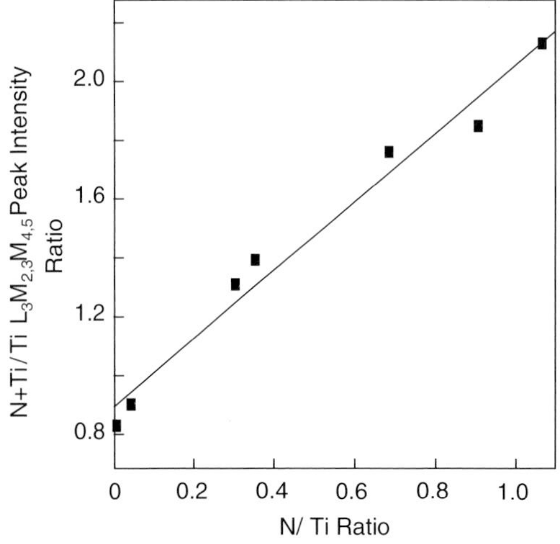


Figure 7.7 Calibration curve to relate the convolution of nitrogen and titanium features to the true N/Ti ratio which has been determined by XPS. (*Source:* reproduced from Baker [1995], with permission).

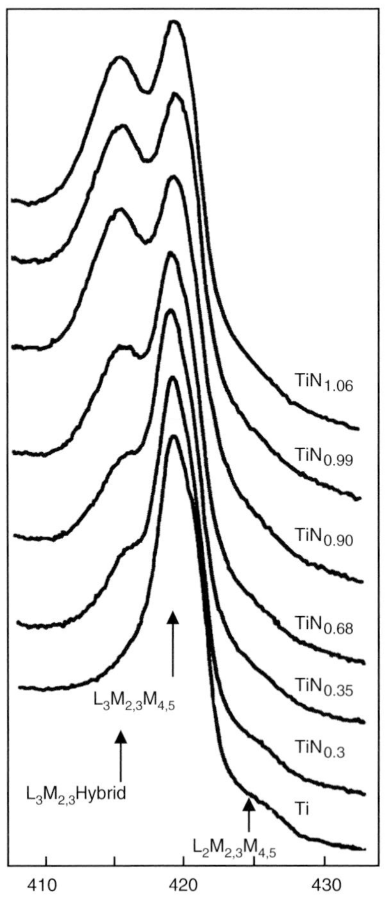


Figure 7.8 $TiL_3M_{2,3}M_{4,5}$ transition for a series of $TiN_x$ coatings. (*Source:* reproduced from Haupt et al. [1994], with permission).

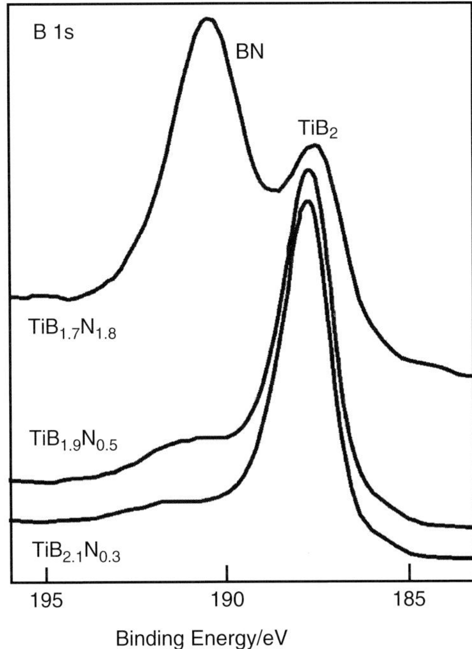


Figure 7.9 B1s XPS spectra from TiN, $TiB_2$, BN, and $TiB_xN_y$. (*Source:* reproduced from Baker et al. [1997], with permission).

indicates that there is no carbide contribution at ca. 282 eV, Figure 7.10., There is, however, a large component at a binding energy between that of the carbide and pure carbon. This is a result of carbon substitution into the $TiB_2$ phase instead of forming the two-phase carbide and boride structure. The weak feature at ca. 287 eV is thought to be a result of oxidation at the end of the deposition process.

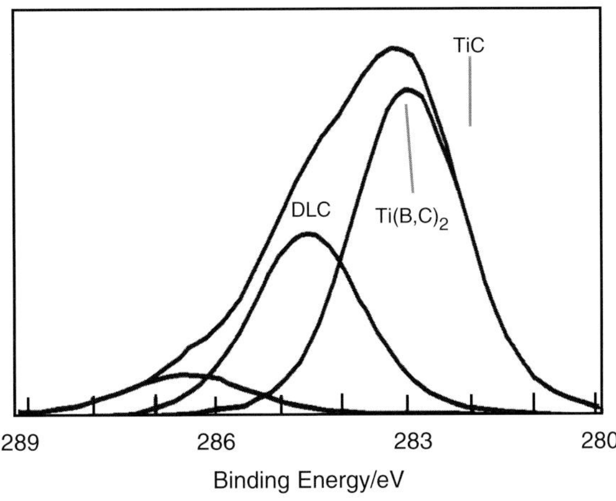


Figure 7.10 The C 1s spectra from titanium borocarbide ($TiB_{1.4}C_{1.2}$). (*Source:* reproduced from Baker et al. [1999], with permission).

At lower carbon concentrations, this occurs in preference to the formation of the DLC phase. These two examples serve to indicate the power of XPS to identify and quantify surface phases, it is then possible to relate such observations back to the phase diagram to establish whether equilibrium or metastable phases exist in the coating.

## 7.3 Corrosion Science

There are two areas within corrosion science in which electron spectroscopy has had a dramatic impact; the interaction of a metal surface with its environment, perhaps to form a passivating overlayer, and the breakdown of the surface film by a localised phenomenon such as pitting. The former is readily studied by XPS; where the ability to separate the spectrum of the underlying metal from that of its oxide film enables the definitive identification of passivating films on alloys. It also enables the thickness of very thin films to be estimated using the Beer-Lambert equation of Section 4.2.1. However, although XPS provides valuable information concerning the composition and perhaps growth kinetics of such films, their breakdown leading to major environmental degradation of the metal, is a localised phenomenon requiring high spatial resolution surface analytical methods, i.e. sub-micron scanning Auger microscopy. The provision of compositional depth profiles of corrosion films is a standard requirement and may be achieved by using argon ion bombardment in conjunction with either AES or XPS. In the case of very thin films ($<4$ nm) angle resolved XPS (ARXPS) can be very informative.

Although in some cases the deconvolution of the metallic and cationic spectral information is straightforward using peak fitting, there are cases of great importance to the corrosion scientist where a major research effort is required to unravel the complexities of two or more valence states along with loss features present in the spectrum. The transition metals in general have presented difficulties. In the case of iron-based alloys, the need is to be able to distinguish Fe(II) from Fe(III) confidently. In the case of the Fe(II) state, these features give rise to a broadening on the lower binding energy side of the valley between the $2p_{3/2}$ and $2p_{1/2}$ spin orbit spitting components, and for Fe(III) on the higher energy side. However, it is the position of the shake-up satellites in the valley between the $2p_{3/2}$ and $2p_{1/2}$ components that give the most reliable information.

The curve fitted spectra of Figure 7.11 show the relative positions Fe(0) and Fe(III) components and the satellites. In the case of a mixed phase, Fe(II) and Fe(II), together with Fe(0), if the film is very thin, must be considered, preferably for both 2p peaks as a check for self-consistency. The backgrounds of the individual singlets will also vary with depth distribution. It is only by using

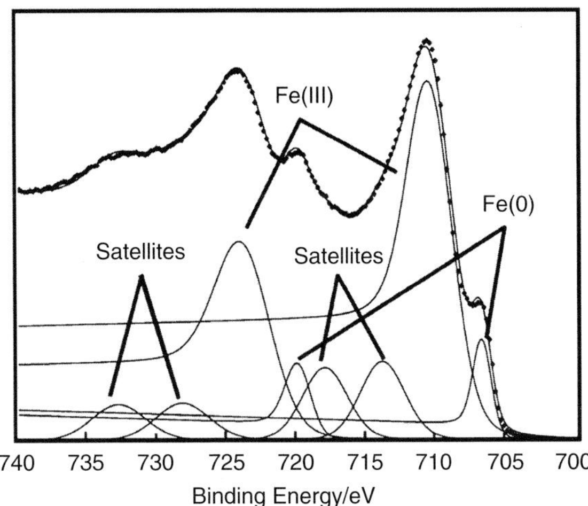


Figure 7.11 The Fe 2p XPS spectrum from an air-exposed steel. Peak fitting has been carried out over the entire Fe 2p window and individual backgrounds with constant tails added to each of the Fe(0) and Fe(III) components. (*Source:* reproduced from Tardio et al. [2015], with permission).

sophisticated computer curve fitting routines in combination with extensive knowledge of peak shape and relative intensities, that this has become possible.

In the case of copper, the shakeup satellites provide a clear route to the identification of copper (II), as seen in Figure 3.10, but the separation of metallic copper and the monovalent ion requires the use of the X-ray induced Auger transition. Information relating to satellite intensity and separation is well documented for all elements of interest to the corrosion scientist although, in some cases, it is necessary to consider the X-ray induced Auger peaks in the XPS spectrum as well. These are particularly informative for magnesium, copper, and zinc, where the photoelectron spectra alone do not provide unambiguous chemical state information.

Electron spectroscopy can also be used to follow the very early stages of oxidation of an alloy, as illustrated by the following example of a U-Mo alloy exposed to the extremely low partial pressure of oxygen in the UHV environment of the analysis system. Considering the Auger data of Figure 7.12a it can be seen that the O *KLL* intensity reaches a plateau at approximately 7 Langmuir (L[1]) exposure of oxygen, equivalent to an exposure to the ambient of <<0.1 μs! The spectra of Figure 7.12b were extracted from a 0–800 eV AES

1 1 L = $10^{-6}$ Torr s, 1 Torr = 1/760 atm = 133.3 Pa. 1 L leads to a coverage of about one monolayer of the adsorbed gas molecules on the surface (assuming that every gas molecule that hits the surface is adsorbed on it). The Langmuir is a convenient unit to use when studying gas adsorption.

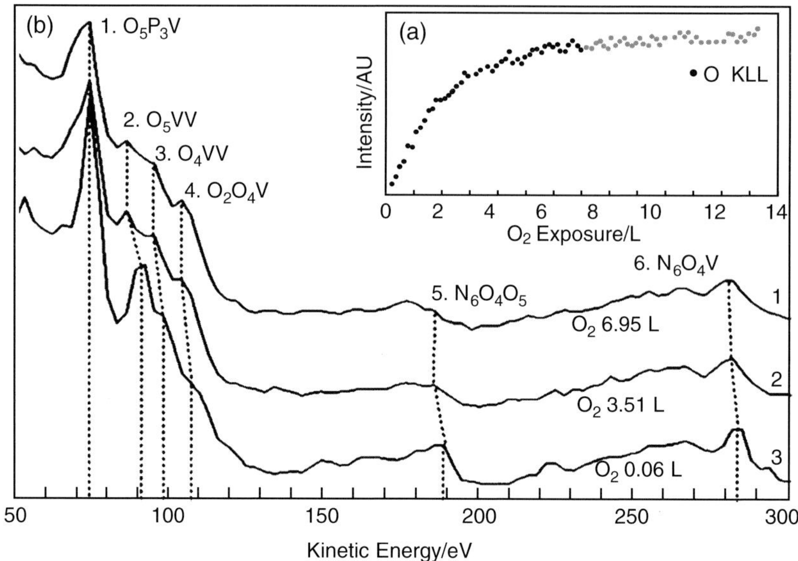


Figure 7.12 Oxidation of a U-Mo alloy at very low oxygen partial pressures. (a) O *KLL* intensity on exposure from 0 to 14 l. (b) Auger spectra showing chemical shift on oxidation at 3.5 and 6.9 l exposure.

Table 7.2 The assignment low-energy uranium Auger peaks of Figure 7.19, comparison of calculated and experimental peak positions and chemical shifts observed on oxidation at 6.9 l.

| Peak | Assignment | $E_{calc}$/eV | $E_{obs}$/eV | Peak shift on oxidation/ΔeV |
|---|---|---|---|---|
| 1 | $O_5P_3V$ | 73.4 | 74.1 | 0.0 |
| 2 | $O_5VV$ | 86.2 | 90.6 | 4.5 |
| 3 | $O_4VV$ | 94.8 | 97.6 | 2.6 |
| 4 | $O_4VV$ | 107.2 | 107.1 | 3.1 |
| 5 | $N_6O_4O_5$ | 184.4 | 187.9 | 3.2 |
| 6 | $N_6O_4V$ | 281.4 | 282.5 | 1.9 |

survey spectrum recorded at a relatively low spectral resolution in order to acquire the data rapidly.

The kinetic energies of the uranium Auger transitions, together with the chemical shifts that occurs on oxidation are given in Table 7.2. The calculated values use the approach of Section 1.4.

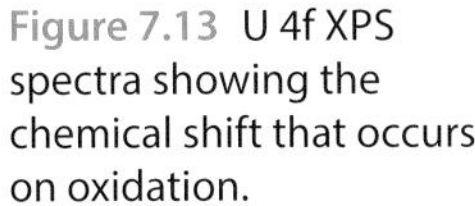

Figure 7.13 U 4f XPS spectra showing the chemical shift that occurs on oxidation.

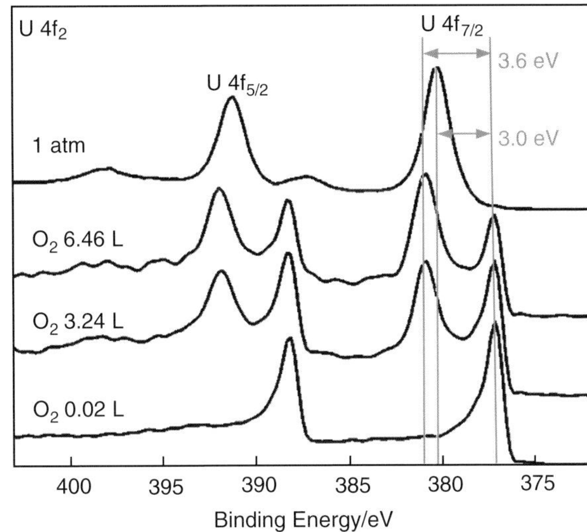


The chemical shifts of the Auger transition of Figure 7.12 are significant (2.0–4.5 eV) and can be clearly seen from the spectra even though they were recorded at a relatively low resolution, consistent with the acquisition of a survey spectrum. Examination of a complementary set of XPS data shows a clear chemical shift between metallic and oxidised components, as shown in Figure 7.13.

The initial shift in both U 4f components is 3.6 eV for the very early stages of oxidation but, on exposure to air, the shift reduces to 3.0 eV consistent with the formation of stoichiometric $UO_2$, by which point the oxide is thick enough that the metallic uranium component of the U 4f spectrum is completely obliterated.

A little used, but nevertheless valuable, method by which surface analysis can assist in the determination of electrochemical history, is by the monitoring of cations or anions absorbed from aqueous solutions. For instance, if a metal electrode is polarised cathodically in $MgCl_2$ solution, it will preferentially adsorb cations on the metal surface, seen in the electron spectrum (XPS or AES) as an excess of magnesium to chlorine. If the electrode has been polarised anodically the reverse is true. This approach is useful in establishing if a pit, or other corrosion site, is active or benign. It can also be used to assess the potential distribution around an active pit. An experiment of the latter type is described in Figure 7.14. The broken line and right-hand axis indicate the anion/cation ratio determined on large electrodes as a function of electrode potential for a constant charge passed, plotted against the predicted potential distribution for small pits. Microanalysis of three pits of about 2 μm in diameter by Auger spectroscopy allows the electrode potential distribution around a pit

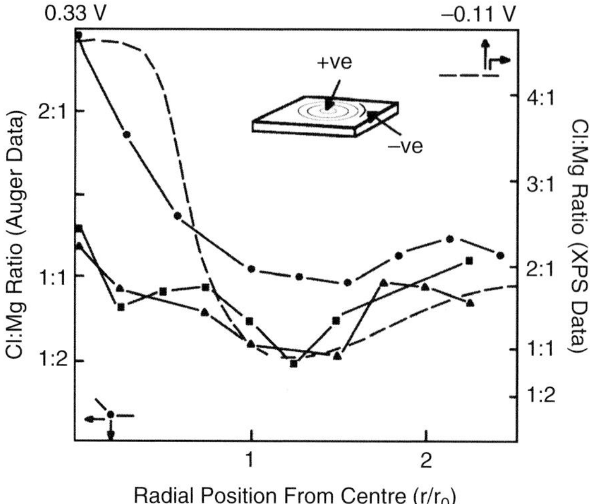


Figure 7.14 Electrode potential around a pit. The dashed line represents the predicted potential distribution obtained from large area (XPS) analysis and consideration of pit geometry (upper and RHS axes). The data points represent Auger analyses made on very fine pits (lower and LHS axes). (*Source:* reproduced from Castle [1986], with permission).

to be determined experimentally, and show excellent agreement with that predicted by the broken line (the combination of XPS and theory), indicating an active anodic centre to the pit surrounded by a cathodic halo as illustrated in the schematic inset of Figure 7.14.

Such pitting may be related to microstructural features of the alloy, and the addition of X-ray analysis facilities to a scanning Auger microscope enables features such as inclusions to be identified and imaged at the same time as the surface phases. In Figure 7.15 a combination of Auger and X-ray maps indicate that a pit, identified in the scanning electron microscope (SEM) image is shown to be active (by the concentration of chlorine in the surface analysis), and associated with a CuS inclusion group, identified by energy dispersive X-ray spectroscopy (EDX). Figure 7.15 also summarizes the other information of value in corrosion studies, the chemical state information from XPS, and the compositional change within the very thin passive film obtained by sputter depth profiling.

The versatility of a scanning Auger microscope fitted with an energy dispersive X-ray detector has been illustrated by studies of the initiation of pitting corrosion in a stainless steel at oxide and manganese sulphide inclusions. In order to follow the corrosion process as a function of time it is necessary to re-establish the sample repeatedly in the Auger system for analysis, following exposure to the corrosive environment. This is best achieved by the judicious use of microhardness indents around the inclusion group of interest. Although

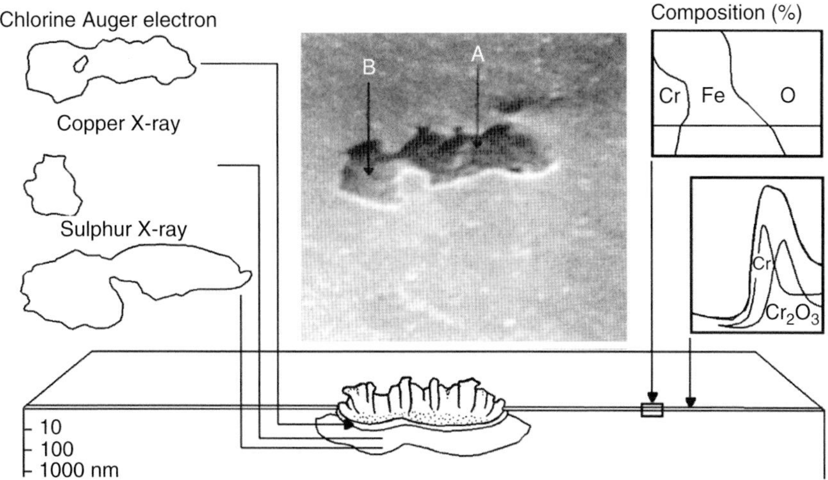


Figure 7.15 A combination of analytical methods has been used to define the corrosion processes occurring around the inclusion group identified as sulphide containing by EDX (middle left). The passive film is characterised by Auger depth profiling and XPS, (upper right). The chlorine Auger map indicates the active area which corresponds to the pit observed in the SEM image (which is 6 μm wide). (*Source:* reproduced from Castle [1986], with permission).

inclusions in steels will have good contrast in optical microscopy this is not the case with electron microscopy and until corrosion features are evident in the secondary electron image relocation of the sample is a very uncertain process. Hence the need for localised fiducial markings such as the microhardness indents.

A typical oxide inclusion is shown in Figure 7.16a following one day of exposure to acidified sodium chloride solution, whilst Figure 7.16b is the same inclusion after 63 days exposure.

A complementary set of SAM and EDX images, taken after 30 days exposure is presented in Figure 7.17.

The micrographs of Figure 7.16 indicate that pitting has initiated, as expected, at the oxide inclusion/metal boundary, and the X-ray images of Figure 7.17 show that the inclusion is a mixture of Mn/Ti/Al oxides. The SEM images show the presence of corrosion deposits adjacent to the inclusion, the surfaces of which are enriched in oxygen, chlorine, silicon, titanium, and managanese, as indicated by the Auger images. Clearly there has been a reduction of pH within the crevice which has developed at the oxide/metal interface which has led to the partial dissolution of the inclusion and the deposition of titanium and manganese ions on the adjacent regions. These zones have also been decorated by silicon which is thought be the result of the dissolution of a soluble silicate which is a minor component of the oxide inclusion.

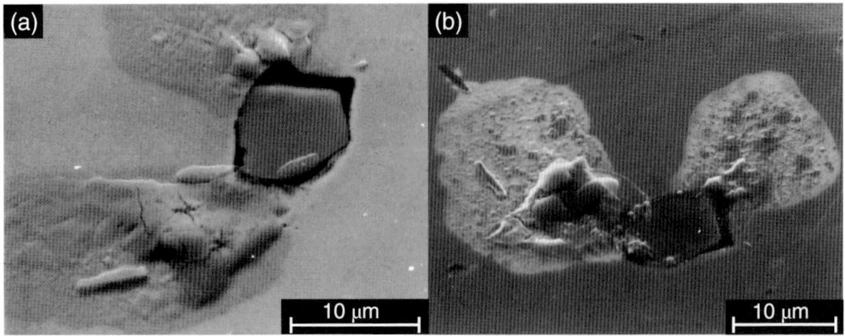


Figure 7.16 A typical oxide inclusion (a) following one day of exposure to acidified sodium chloride solution and (b) after 63 days exposure. (*Source:* reproduced from Baker and Castle [1992], with permission).

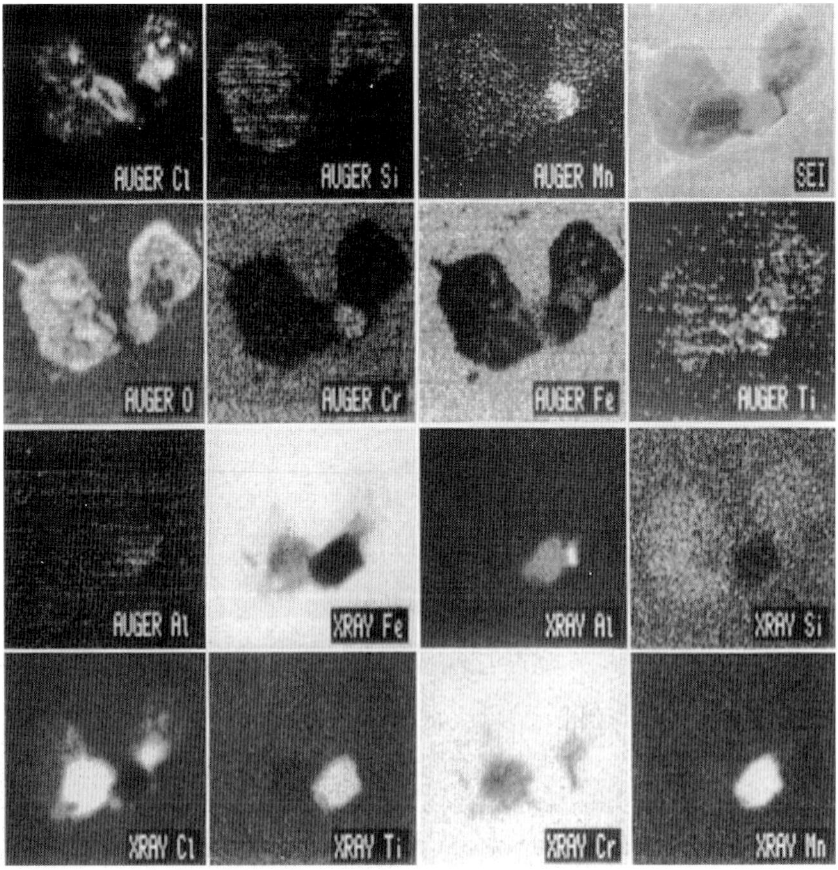

Figure 7.17 A complementary set of SAM and EDX images, taken after 30 days exposure to acidified sodium chloride solution. (*Source:* reproduced from Baker and Castle [1992], with permission).

A similar approach has been used to study the very early (after ten seconds exposure to saline solution) dissolution of manganese sulphide inclusions in steel. The model of Figure 7.18, indicates the reactions and transport processes at, and around, several pits associated with such inclusions and was deduced from Auger analyses.

The benefits of this approach to the investigation of corrosion phenomena is that reactants and products associated with the chemical reactions can be identified *in situ* and reaction mechanisms proposed on the basis of analytical chemistry results rather than inferred from electrochemical measurements and morphological observations. Although the methodology was initially

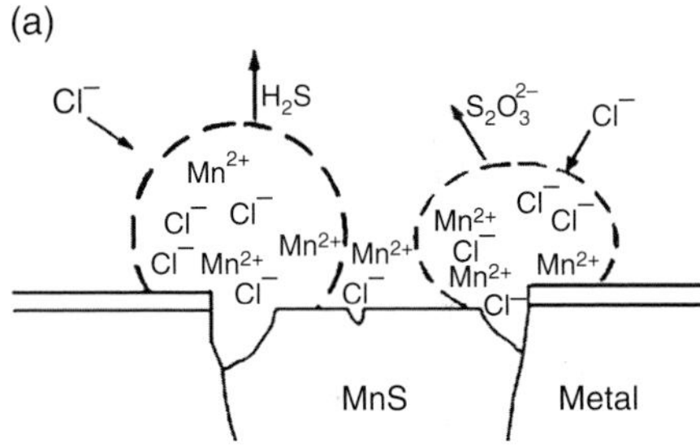


Initial Stages of MnS Dissolution

(a) Anodic nature of dissolving inclusion and loss of S-containing anions from electrolytic system (formation of $H_2S$, elemental S as corrosion products) attracts a high local concentration of $Cl^-$ to the site.
(b) Largest concentration of $Mn^{2+}$ and $Cl^-$ accumulating in the bulk solution above the small cavity.
(c) Exposed bare metal repassivates

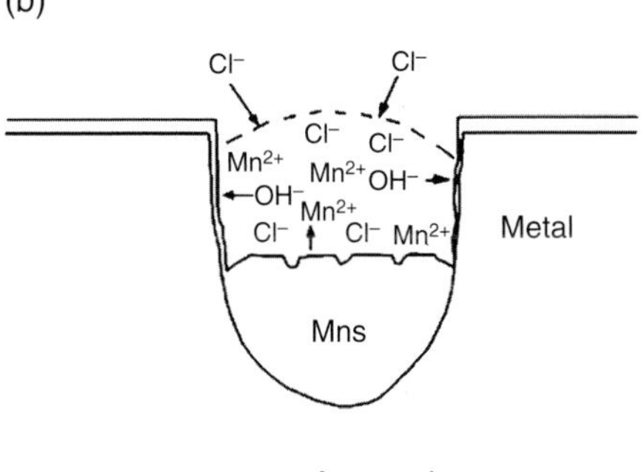


Partial MnS Dissolution

(a) Further dissolution causes an increasing. $Mn^{2+}$ and $Cl^-$ concentration within the cavity.
(b) Uni-directional diffusion restricts the transport of corrosion products away from the cavity.
(c) Exposed cavity wall repassivates.

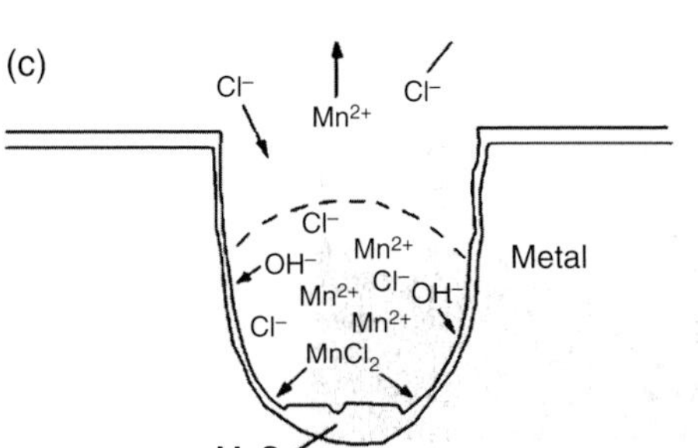


Further MnS Dissolution and Stabilisation of Pit Growth

(a) Critical concentration of $Mn^{2+}$ and $Cl^-$ attained for precipitation of $MnCl_2$ salt film.
(b) The salt film prevents repassivation and provides conditions favouring stabilised metallic corrosion.

Figure 7.18 Model of the early dissolution of manganese sulphide inclusions in steel. (*Source:* reproduced from Baker and Castle [1993], with permission).

developed for the study of pitting corrosion in stainless steels it has now been employed to good effect to unravel the microchemistry of pitting corrosion associated with inclusions in aluminium alloys and beryllium.

## 7.4 Ceramics

In the field of ceramics, it has been XPS that has proved the more useful of the two techniques, with problems of sample charging limiting the number of investigations carried out by AES. There are several possible approaches to the analysis of insulators by AES including the use of a low-energy beam of argon ions, grazing incidence of the incident electron beam and the use of a metallic mask to effectively 'bleed off' surface charge. The choice of method to use for the analysis of insulators by AES is largely a matter of trial and error but the first two methods are illustrated by examples of the analysis of bulk ceramics, and the latter in the catalysis example below.

The Auger spectra and image of Figure 7.19 are from a polished silicon nitride sample. The specimen has not been sintered to full density and consequently the voids are identified by the $SiO_2$ present at the surface of the SiN

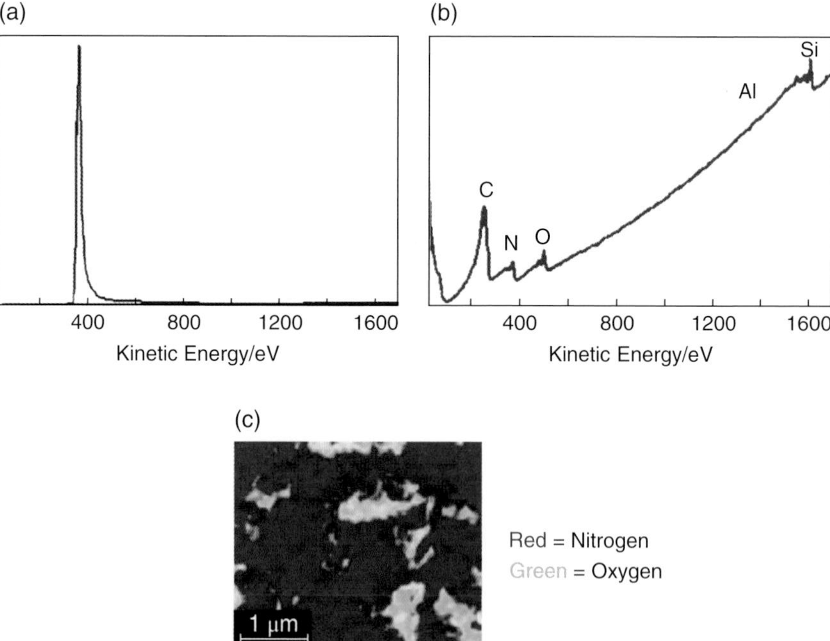


Figure 7.19 Auger analysis of an insulating oxidised silicon nitride sample. (a) Spectrum recorded with no attempt at charge compensation, (b) application of a low energy (20 eV) argon ion beam, (c) scanning Auger microscopy identifying the oxidised and unoxidised regions of the sample.

particles prior to consolidation and heat treatment. Figure 7.19a shows the spectrum recorded without any form of charge compensation. The electrostatic charging of the surface has distorted the spectrum to such an extent that a single broad peak at around 400 eV is the only feature recorded.

The efficacy of a low-energy ion beam as a method of charge compensation is seen in Figure 7.19b where an excellent Auger spectrum of the sample is presented. The intense C *KLL* peak at ca. 270 eV is a result of polishing media in the porosity of the sample segregating to surface within the UHV of the scanning Auger microscope. The uniformity of charge compensation over a $4 \times 4\,\mu m^2$ area is seen in these Auger images, where the polished SiN and the pores identified by the presence of $SiO_2$ are clearly delineated.

If an insulating sample cannot be exposed to low-energy ions, or if the method is not successful for charge compensation, the use of a low-energy electron beam at grazing incidence can be used. The principle here is that as the beam is at grazing incidence the area on the sample irradiated will be large and thus the electrons emitted per unit area will be lower, this, coupled with a low-energy beam, often enable the emitted electrons to be compensated by any leakage current from earth. The spectra of Figure 7.20 are from a thick sample of BeO, previously ion etched clean in the spectrometer, a 3 keV beam, at an angle of 85° to the surface normal, was used to record the spectra.

In this manner, Auger spectroscopy is used in the fields of catalysis and mineralogy and one can be sure that this trend will continue. In this section we shall consider the role that electron spectroscopy has to play in the analysis of catalysis samples and naturally occurring minerals.

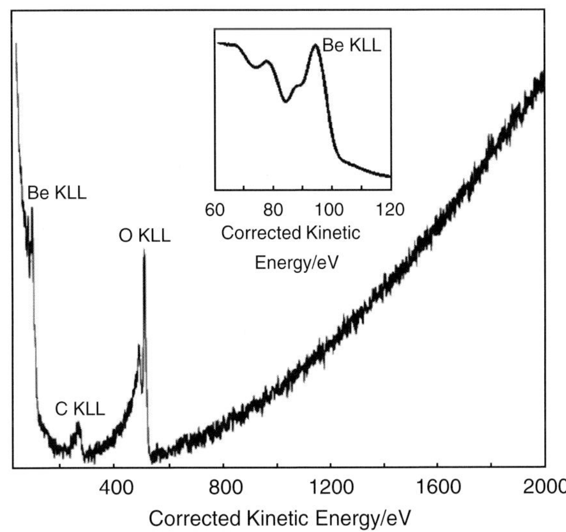


Figure 7.20 The AES spectrum of BeO, recorded at grazing incidence with a 3 keV electron beam. Inset is the high resolution Be *KLL* transition.

In the application of XPS to catalysis studies, there appear to be three areas of endeavour:

1) The furthering of the basic science of heterogeneous catalysis has relied greatly on the pure surface science approach, i.e. the preparation of metal or inorganic single crystals with a pre-defined crystal orientation which is then exposed to very small quantities of reactant(s) in the gas phase. In this manner, the reaction occurring on the crystal surface can be followed in a stepwise pattern, the modification of substrate or adsorbate being apparent from the electron spectrum. Such experiments are often carried out in conjunction with low-energy electron diffraction (LEED) which yields information concerning surface crystallography. With the advent of commercial, near-ambient pressure XPS systems the prognosis for major advances in this area in the near future seems extremely good.
2) The activity of a supported catalyst is frequently a function of the level of dispersion of the metal or oxide on the support medium. The size of such supported crystallites can sometimes be estimated from the intensity if the appropriate XPS peak ratio. However, this does require assumptions regarding particle shape and the most rewarding studies appear to be those which combine XPS data with a TEM study.
3) The area in which surface analysis has made the most spectacular impact is in the identification of catalyst poisons, and other trouble-shooting investigations.

An example of the use of AES in comparing fresh and spent catalyst is presented in Figure 7.21. The two spectra were obtained from a 0.5% Pd catalyst supported on $Al_2O_3$ with chromium and molybdenum additions as promoters. Comparison of the Auger spectra from the two samples indicates that de-activation is associated with a large concentration of iron attenuating the Al, Pd, Cr, and Mo signals present on the clean surface.

Thus, the poor performance of this material could be associated with an iron contaminant, probably emanating from steel pipework or reaction vessel, masking the highly active palladium atoms as well as the promoter atoms, and greatly reducing the catalytic activity of the material.

The surface analysis of naturally occurring minerals, although straightforward in principle, provides many problems in practice. Whilst the provision of an elemental surface analysis by XPS is straightforward, extracting the required level of chemical, or indeed structural, information can be difficult. There are two well-established methods that can be applied to extend the level of information available from XPS by extending the analysis from the usual core level peaks to either valence band (VB) studies or the measurement of the Auger parameter.

An elegant example of the use of the VB is in the study of $TiO_2$, which occurs in two crystal forms; rutile and anatase. $TiO_2$ is used in many applications and

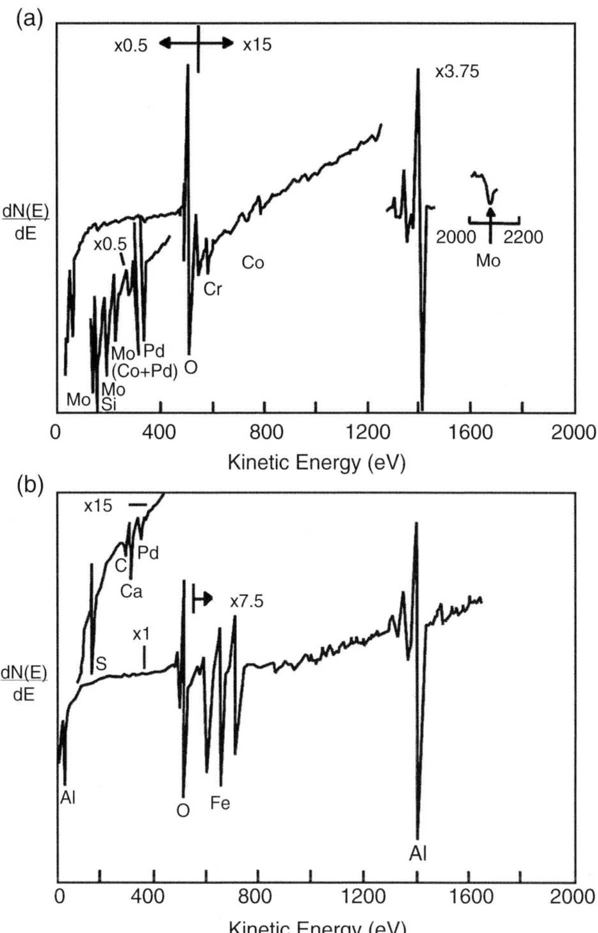


Figure 7.21 AES of fresh (a) and spent (b) alumina-supported 0.5% Pd catalyst. (*Source:* reproduced from Bhasin [1975], with permission).

potential applications including white pigmentation for paints, and photocatalytic purposes such as solar cells and water purification. The XPS Ti 2p spectra of both forms of $TiO_2$ are identical whereas the XPS VB of the two phases are significantly different, as shown in Figure 7.22.

There are clear differences between anatase and rutile which can be defined with three separate components to provide a fingerprint shape for each $TiO_2$. In the case of a mixed phase sample, the surface ratio of the two phases (which is directly related to photocatalytic activity, the more anatase the better) can be readily determined by fitting using the models of Figure 7.22.

In the case of aluminosilicate minerals, along with some other naturally occurring materials, there are two additional problems involved. First,

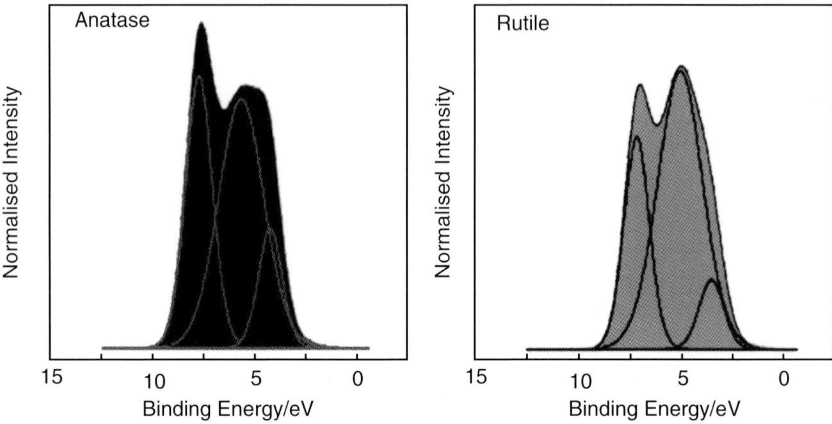


Figure 7.22 The XPS valence bands (VBs) of the anatase and rutile forms of $TiO_2$. (*Source:* Reproduced from Breeson et al. [2017], with permission).

electrostatic charging often means that the confidence level on the charge corrected peak position is greater than the spectral window containing the chemical information! Secondly, the peaks often have a very poor photoelectron cross section (this is particularly true of Si 2p, and Al 2p). The means by which these problems have been overcome are closely related to each other and involve measuring the associated Auger transitions (i.e. Si *KLL* and Al *KLL* for aluminosilicates), which with a non-monochromatic source is sometimes accessible using the bremsstrahlung component of the X-ray source.

The problem of sample charging may be overcome by reporting the separation of two peaks rather than the absolute binding energy of a photoelectron transmission. In some elements the Auger chemical shift is equal to and sometimes greater than the XPS chemical shift, and the potential exists for extracting chemical information via the Auger parameter (α) defined in Section 3.3.4, as:

$$\alpha = E_K + E_B$$

where $E_B$ is the binding energy of the XPS peak (e.g. 1s or 2p) and $E_K$ is the kinetic energy of the attendant Auger peak (e.g. $KL_{2,3}L_{2,3}$). In the case of aluminium and silicon the $KL_{2,3}L_{2,3}$ Auger transition is not directly accessible, although the bremsstrahlung radiation from a non-monochromated X-ray source is able to eject sufficient Si 1s electrons to produce a measurable Si $KL_{2,3}L_{2,3}$ peak. The silicon Auger parameter calculated in this way is independent of sample charging but strongly dependent on both molecular and crystalline structure.

There still exists the problem of poor photoelectron cross section of the Al 2p and Si 2p levels and the only way to overcome this is to use a higher energy X-ray anode able to excite 1s core levels of these elements. Various possibilities exist

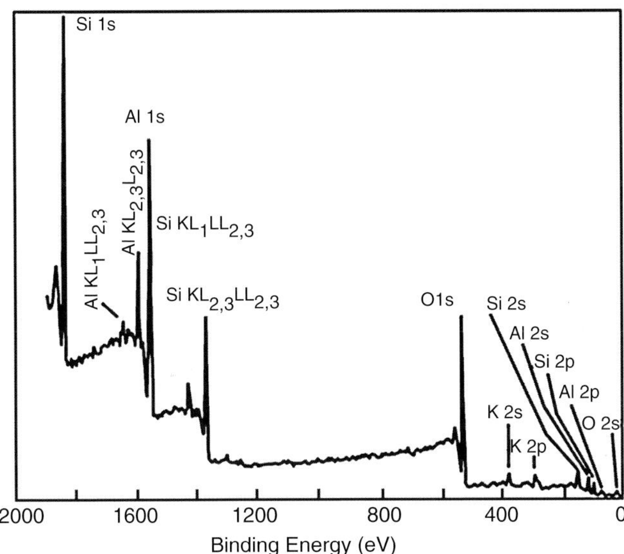


Figure 7.23 XPS survey spectrum of muscovite mica recorded with monochromatic Ag Lα radiation.

including Zr Lα and Ti Kα, but the ones which represent the best combination of sensitivity and resolution are monochromated Ag Lα and Cr Kα. Using such a source, it is possible to record 1s-*KLL* Auger parameters; a spectrum obtained with a monochromated Ag Lα high-energy source of this type is shown in Figure 7.23.

One final application of XPS in the analysis of ceramics is that of photoelectron forward scattering, also referred to as X-ray photoelectron diffraction (XPD). If the sample under consideration is a single crystal, photoelectrons emitted from sub-surface planes will be scattered by surrounding atoms giving rise to an angular distribution of the emitting site concerned.

This type of angular modulation is quite different to that utilised in the non-destructive depth profiling of materials described in Chapter 4. Indeed, for successful forward scattering, the crystal surface should be extremely clean, preferably prepared *in vacuo* by, for instance, cleaving. This procedure has been used with a high degree of success for the investigation of complex minerals, and also semiconductors such as gallium arsenide. It would seem that such a method offers much promise as a way of ensuring the epitaxial orientation of very thin layers.

## 7.5 Microelectronics and Semiconductor Materials

The trends in the development of semiconductor devices continue to be towards both increasing transistor density and higher operating frequencies. The consequences of this are that individual structures on the chip are becoming smaller and thinner and the size of defects which cause chip failure (critical defects) are also becoming smaller. The fact that these devices are built at or near the surface of a silicon wafer means that surface analysis techniques are commonly used in the semiconductor industry.

The spatial resolution available from Auger electron spectroscopy makes it a useful tool for the analysis of device structures. The depth profiling capabilities of the technique are also used to assess the quality of the layered structures that form part of modern semiconductor devices. XPS is also used for its ability to determine the chemical states of the, sometimes complex, materials which are produced.

The trend towards thinner layers means that dielectric films are becoming amenable to study using ARXPS. The need for XPS is expected to increase when silicon dioxide is eventually replaced by more exotic materials having a much higher dielectric constant. With these layers, it will become important to ascertain the chemistry not only of the layer but also of the interface between the layer and the silicon or silicon dioxide substrate. ARXPS will be able to do this without the need to sputter the material and risk the alteration of the chemical states present.

### 7.5.1 Mapping Semiconductor Devices Using AES

The complexity of modern semiconductor fabrication methods is such that it is necessary to check from time to time that the intended structures are being produced. Scanning Auger microscopy measurements are very useful for this. Figure 7.24 is an example of a set of such measurements. It is usual to start the analysis by producing a secondary electron micrograph, this provides an indication of the topography of the sample and helps with the selection of the area to be analysed by Auger. As well as the SEM image, Figure 7.24 shows scanning Auger images from Si, O, N, Al, and W. Clearly the features in these images are well defined and uniform in size and shape. There is no suggestion of diffusion or migration of any of the elements into regions where they should not be.

The analysis that resulted in Figure 7.24 was a challenging one. The images were derived from a cross section through the surface region of the wafer. The sample was prepared using a focused ion beam (FIB), in the manner illustrated in Figure 4.30. Once the sample was removed from the wafer it was laid on a graphite support and transferred to the Auger instrument. As can be seen in Figure 7.24 much of the sample consisted of the insulating materials silicon dioxide and silicon nitride, both of which are good insulators. For this reason, the sample had to be very thin (~100 nm) to avoid sample charging, see Section 6.6.2. If the sample were to charge then the position of the Auger peaks would shift and the images, including the SEM, would become severely distorted. Other methods of controlling the charging on this sample are unlikely

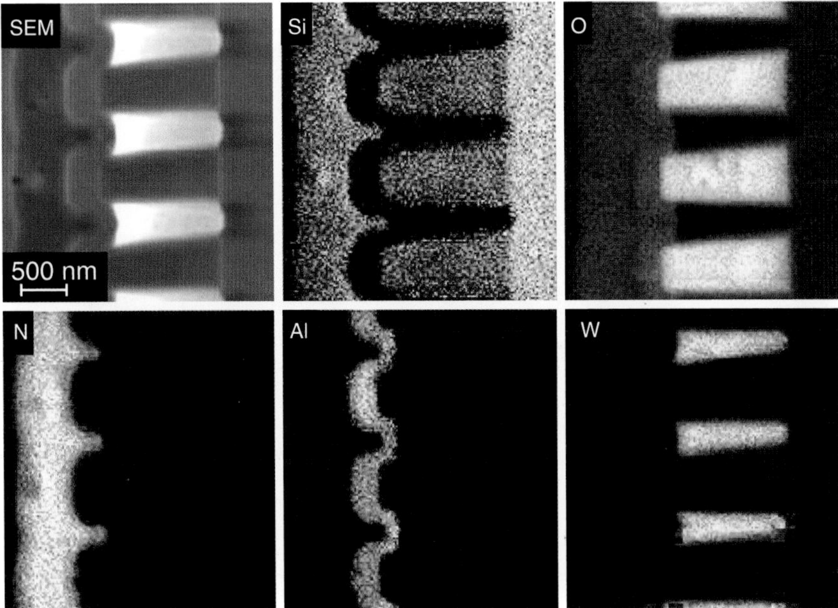


Figure 7.24 The analysis of a complex structure using secondary electron imaging and scanning Auger imaging.

to be effective. Balancing the electron emission with the input beam current would require the use of a low kinetic energy beam. If a low-energy beam were to be used the spot size would be too large to achieve the required image resolution. The use of low-energy ions would not be practical either because it would be very difficult to achieve a sufficiently high-ion current density when using a beam energy that would not cause significant sample sputtering.

A second important role for Auger in device fabrication is in defect review, Figure 7.25 provides a simple example. Figure 7.25a shows the SEM image of a set of TiN-capped metal lines on $SiO_2$. A nearly transparent particle can be seen in the SEM image. Figure 7.25c shows Auger spectra from one of the TiN lines remote from the particle and Figure 7.25d shows a spectrum from the particle. The spectrum from the particle shows intense Al and F peaks with no Ti peak being present in the spectrum. Figure 7.25c shows a small Al peak and intense peaks due to Ti and N (note that it is difficult to resolve the nitrogen peak from one of the Ti peaks). Figure 7.25b is an overlay of two scanning Auger images, Ti in green and Al in red.

The conclusion to be drawn from the data shown in Figure 7.25 is that the contaminating particle is likely to have come from an etch chamber used in the production of the sample. The fluorine observed in the flake is likely to have

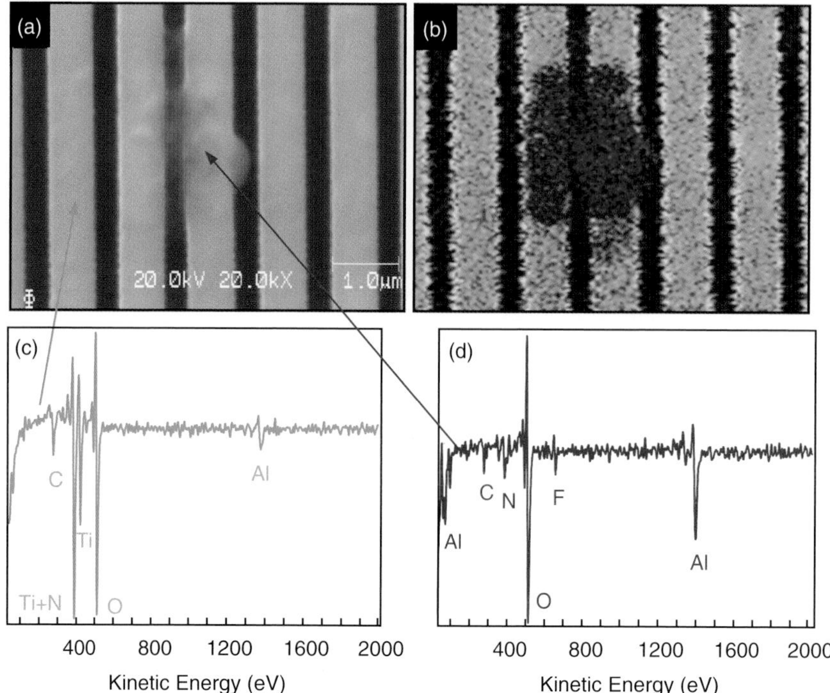


Figure 7.25 (a) SEM image showing a contaminating particle on a set of TiN lines; (b) Scanning Auger maps of Ti (green) and Al (red) acquired from the same area as the SEM; (c) Auger spectrum from an uncontaminated TiN line; (d) Auger spectrum from the particle.

come from the same chamber and was present following one of the periodic cleaning procedures.

Sometimes analysis of particulate contamination is more complex. The sample analysed in Figure 7.26 consists of a set of tungsten lines on silicon dioxide. A contaminating particle was observed on one of the tungsten lines. Analysis of the particle in an analogous way to that shown in Figure 7.25 showed that the surface of the particle only consisted of tungsten, suggesting that the contamination occurred in a step prior to the deposition of the metal. Confirming this required the analysis of the cross section of the particle, achieved with the help of an *ex-situ* FIB. Figure 7.26a is an SEM image of the sample showing a crater produced by the FIB. It can be seen that the crater wall passes through the particle which is shown at higher magnification in Figure 7.26b. The set of Auger maps in Figure 7.26c shows O in red surrounded by elemental Si in green and W in blue. This distribution of species suggests strongly that the original contaminating particle was $SiO_2$ and it was introduced during the deposition of poly-silicon.

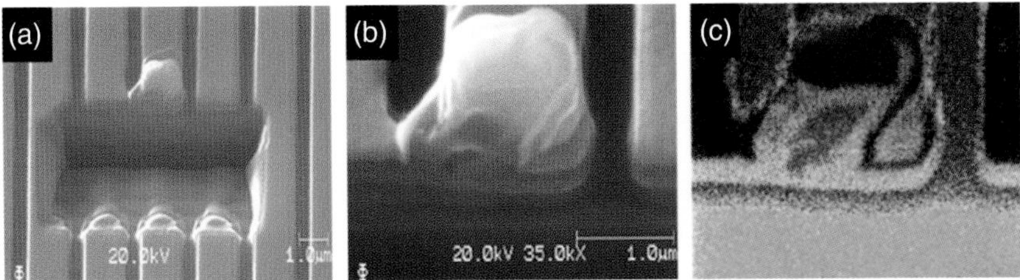


Figure 7.26 (a) SEM image of a FIB cross section through the particle; (b) As (a) but at higher magnification; (c) An overlay of three Auger maps from the cross section.

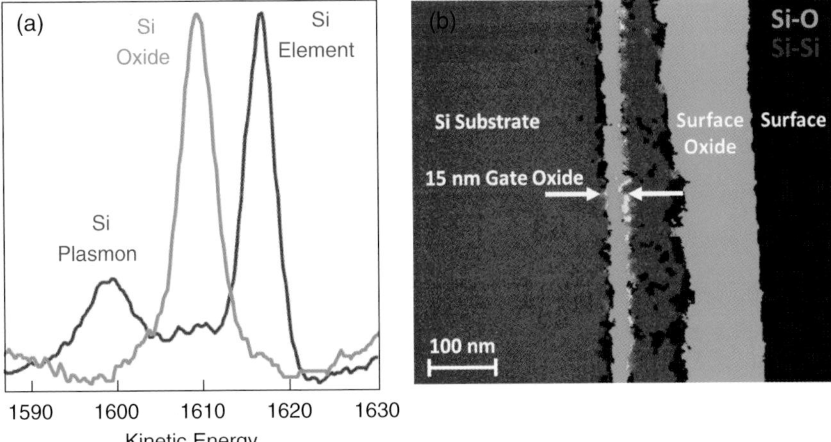


Figure 7.27 (a) Auger spectra from elemental silicon and oxidised silicon showing the chemical shift which occurs in the oxide; (b) The chemical states of silicon (elemental and oxide) can be mapped with high spatial resolution using AES.

Scanning Auger microscopy is capable of more than elemental analysis. If the instrument has reasonable energy resolution, then chemical state maps can be produced. When the separation of Auger peaks is large, as is the case with elemental silicon and its oxide, Figure 7.27a, then mapping the chemical states is straightforward, see Figure 7.27b. The map in Figure 7.27b is from a cross section of a silicon wafer and shows a surface oxide layer and a thin (15 nm) gate oxide layer. In the case of silicon, it is not only the element and its oxide that can be independently mapped. Silicon in its elemental form can be distinguished from silicon in the form of a silicide. For example, the energy difference between the Si *KLL* peaks from the element and from nickel silicide is about 0.7 eV so the materials may be distinguished if the spectrometer is operated with an energy resolution in the region of 0.1%.

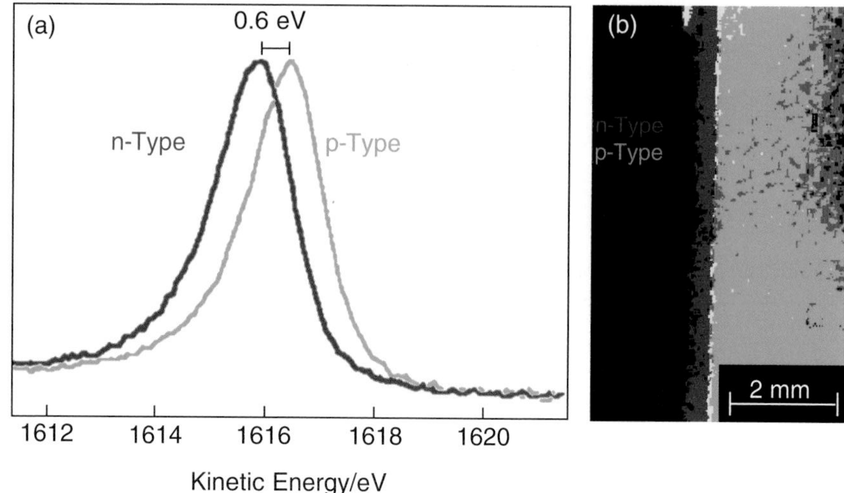


Figure 7.28 (a) Spectra from n-type and p-type silicon show a small shift in peak position (0.6 eV); (b) Overlay of maps derived from n-type and p-type silicon.

Using even better energy resolution, mapping of dopant types becomes possible. Figure 7.28a shows that there is a small energy difference (~0.6 eV) between the Si *KLL* peaks from n-type and p-type silicon. A similar phenomenon is also observed in XPS. The concentration of the dopants in these materials is extremely small, far below the detection limits of Auger electron spectroscopy. Nevertheless, the energy shift is sufficient to allow maps to be produced. As an example, Figure 7.28b shows a map derived from the cross section of a silicon sample implanted with phosphorus in the near-surface region to produce n-type silicon.

### 7.5.2 XPS Failure Analysis of Microelectronic Devices

Although microelectronics is being driven towards smaller and smaller devices, there are often examples where it is desirable to analyse features with dimensions of tens to hundreds of micrometres. Such an example is the attachment of leads to bond pads, where XPS maps and spectra can be gainfully employed to quickly provide analytical information to the failure investigation team. Visual inspection of such a specimen indicates that of the five bond pads observed, three have leads attached, whilst two do not have leads connected Figure 7.29a). The upper of these two (Pt 1) is where a lead has detached and the lower (Pt 2) is one that is not used. The question facing the analyst is whether the failure of the lead attached to the upper bond pad is a result of pre-existing contamination or merely a mechanical failure. The lower bond pad provides a useful benchmark for analysis. Figure 7.29 shows examples of

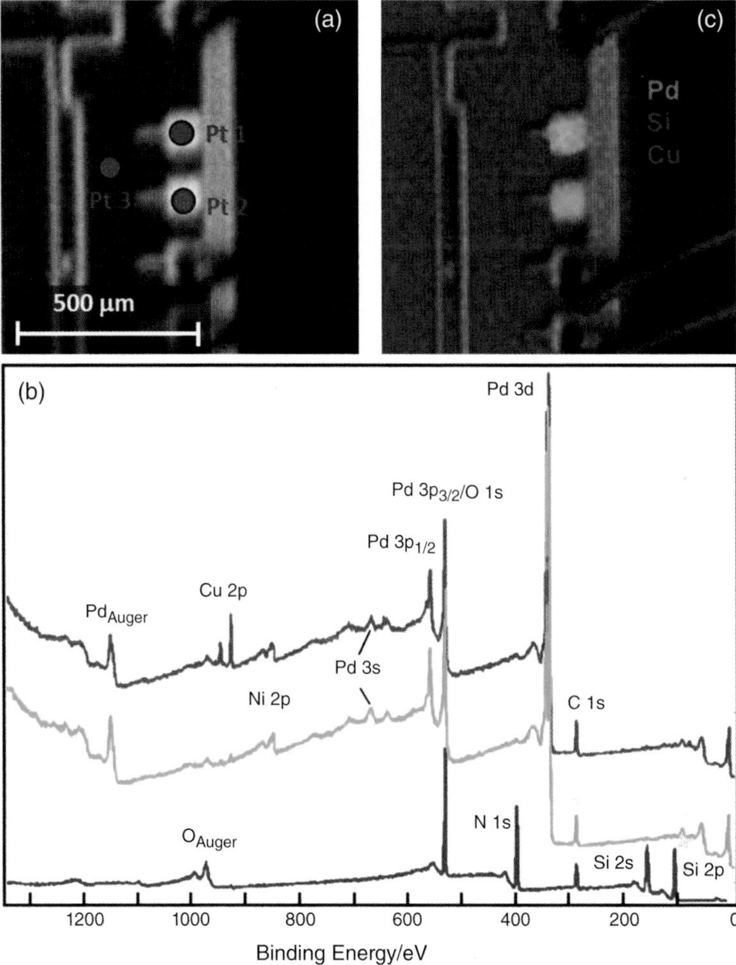


Figure 7.29 Examples of XPS SnapMaps™ and spectra from a bond pad failure investigation on a microelectronics device. (a) optical image showing three points chosen for point analysis, (b) XPS survey spectra from the three points identified in (a), with quantitative data below (Table 7.3), (c) XPS false colour elemental SnapMaps™ of palladium silicon and copper.

Table 7.3 Quantitative surface analyses derived from the three spectra shown in Figure 7.29b.

| | Surface composition/atomic % | | | | | | | |
|---|---|---|---|---|---|---|---|---|
| | C | O | Pd | Ni | Cu | N | Si | Cl |
| **Point 1** | 36.9 | 18.9 | 30.0 | 7.4 | 4.8 | 0.0 | 0.9 | 1.0 |
| **Point 2** | 34.7 | 20.3 | 34.4 | 7.4 | 1.0 | 0.0 | 2.2 | 0.0 |
| **Point 3** | 14.2 | 27.5 | 0.0 | 0.0 | 0.3 | 25.9 | 32.1 | 0.0 |

XPS images from this investigation. The bond pads are the five square features arranged vertically in the centre of the images.

The images were acquired as SnapMaps[2] by monitoring the intensity of the desired XPS peak in the SnapShot mode (see Section 2.9.1), i.e. without scanning the spectrometer, merely relying on dispersion of the electrons at the multichannel detector to provide the spectrum. This is a very rapid approach to spectral acquisition and combined with a short dwell time of the sample under the X-ray beam, as the stage is quickly rastered over the $1 \times 1\,mm^2$ area of interest, allows a full spectrum map to be acquired in a very short time. The images of Figure 7.29 illustrate this point, with the $100 \times 100$ pixel images being acquired in about five minutes each. Figure 7.29a shows an optical image, which enables easy identification of features at the sample surface, showing two bright bond pads (no leads attached) and three less distinct ones which have leads attached. This image can be used for exact placement of a 30 μm X-ray beam to carry out small area XPS analysis of the three regions, identified in the map of Figure 7.29a, by red (1), green (2) and blue (3) solid circles. The spectra obtained from these three regions are shown in Figure 7.29b, and the quantitative data from these three points are presented in Table 7.3.

The analysis of Point 1 (red), shows a significant increase in copper compared to the lower bond pad (Point 2), 4.8 at.% cf. 1.0 at.%. Point 3 represents the silicon oxynitride substrate. False colour XPS images of the Pd 3d (green), Si 2p (red), and Cu 2p3/2 (blue) intensities are shown in Figure 7.29c, the resolution of these elemental images is 16 μm. The copper leads are clearly seen diagonally attached to the bond pads on the right of the field of view, and the silicon image readily delineates the substrate. The upper bond pad (Point 1) shows a slight blue intensity consistent with the observation of the spectrum from this point of copper associated with the palladium. The presence of copper on this bond pad indicates a successful bond between pad and lead resulting in diffusion of a small amount of copper into the pad. Thus, one can conclude that failure is not a result of bond pad contamination prior to joining but failure of a mechanical nature.

## 7.5.3 Depth Profiling of Semiconductor Materials

### 7.5.3.1 Transistor Gate Dielectrics

Modern dielectric layers, which are used in the gate region of transistors, need to be extremely thin to allow rapid and efficient switching of the transistor without the need for high voltages. The thickness of these materials has become similar to the information depth of X-ray photoelectron spectroscopy.

2 SnapMap is a trademark of ThermoFisher Scientific.

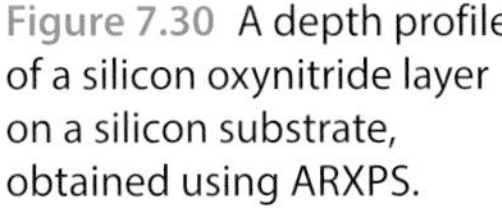

Figure 7.30 A depth profile of a silicon oxynitride layer on a silicon substrate, obtained using ARXPS.

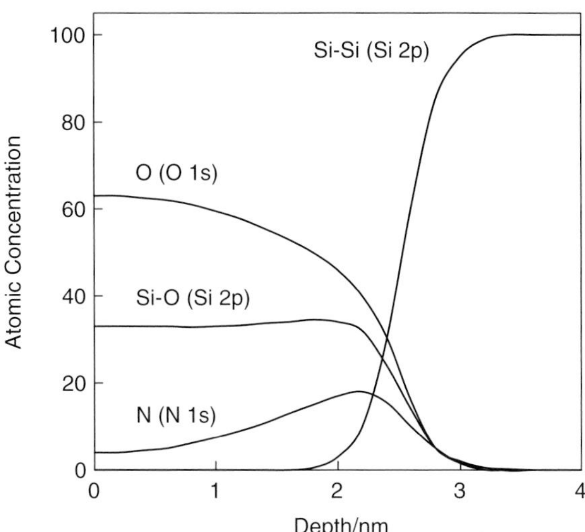


Despite the fact that these layers are so thin, it remains necessary to characterise them, not only in terms of thickness but also in terms of the distribution of elements and their chemical states through the layer. Sputter profiling is of limited use because the escape depth of the photoelectrons is large compared with the layer thickness and so there would be insufficient depth resolution.

There are several examples showing how these ultra-thin layers may be characterised using ARXPS in Chapter 4. Figure 7.30 shows a further example, a depth profile of a silicon oxyitride from which the layer thickness may be calculated along with the nitrogen dose and distribution. Repeating measurements such as this at different points on a wafer and on different wafers gives a good indication of the reproducibility of the fabrication process.

#### 7.5.3.2 Inorganic Chemical State Profiling

Sputter depth profiling using Auger electron spectroscopy is frequently applied to semiconductor materials. The very flat interfaces present in many of the sample types means that extremely good depth resolution can be achieved if all of the instrumental parameters are carefully chosen, as discussed in Chapter 4.

Elemental profiles are performed to ascertain layer and interface purity, layer thickness and migration or diffusion of material from one layer to another. However, chemical state information is frequently required and can be provided by profiling using an Auger spectrometer. Figure 7.31 shows Auger spectra from elemental titanium and in some of its compounds. A profile data set containing titanium spectra can then be fitted with these standard spectra using a least squares method. Figure 7.32 shows the results of this process from a multilayer material on a silicon substrate. For clarity, only the titanium species are shown.

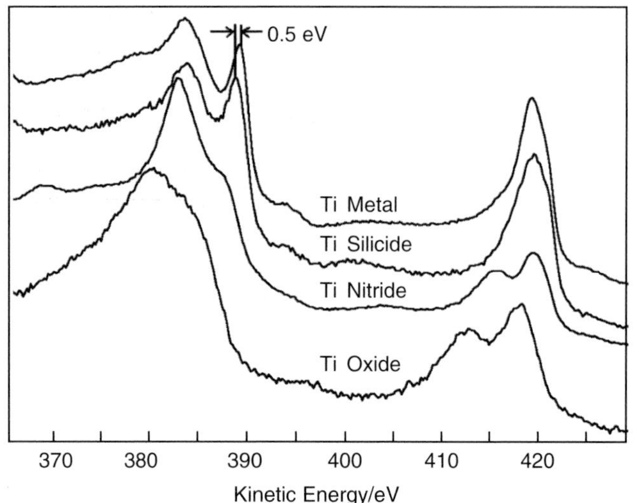


Figure 7.31 Auger spectra from elemental titanium and some of its compounds.

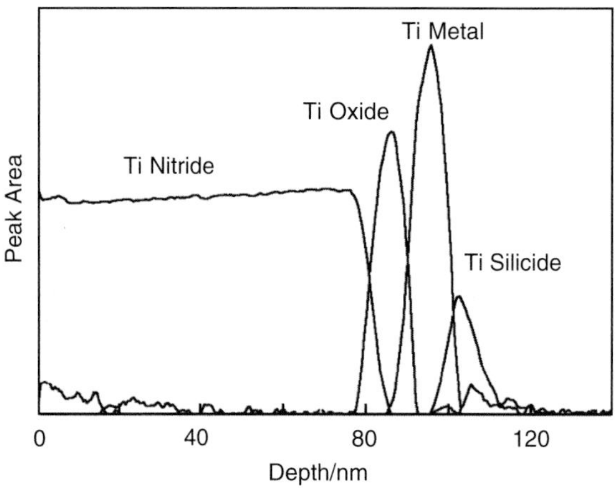


Figure 7.32 Depth profile of titanium nitride/titanium oxide/titanium on silicon. Note the formation of the silicide at the silicon interface.

The method of least squares fitting is sufficiently powerful to distinguish titanium silicide and titanium metal when the spectra are very similar; the silicide spectrum is shifted with respect to the metal spectrum by only 0.5 eV.

#### 7.5.3.3 Organic Semiconductor Profiling

Figure 7.33a is a schematic representation of the structure of an organic field-effect transistor (OFET). As can be seen, the region of the OFET between the source and the drain consists of two layers, the organic semiconductor and the silicon dioxide insulator, on a silicon substrate. In this example, the organic

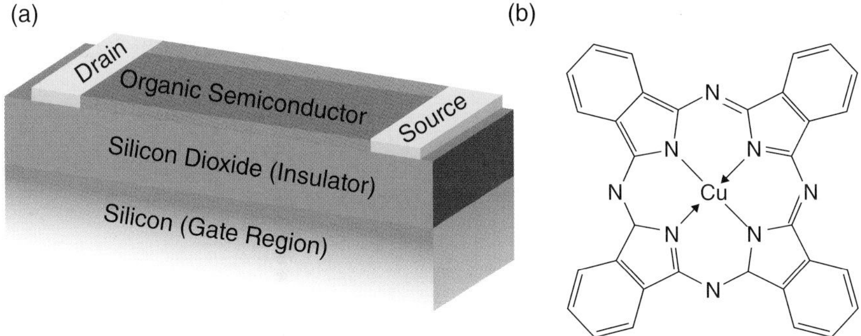


Figure 7.33 (a) Schematic diagram showing the structure of an organic field-effect transistor (OFET); (b) The structure of the organic semiconductor, copper (II) phthalocyanine (CuPC), used in this example.

material is copper (II) phthalocyanine (CuPC) whose structure is shown in Figure 7.33b and which is a p-type semiconductor. The electrical properties of the device depend upon the chemistry of the layers and that of the two interfaces. In order to confirm that the correct chemistry is present throughout the layers it is necessary to perform a sputter depth profile using XPS and to do this in such a way that the chemistry is not altered by the sputtering process.

To minimise the risk of damage to the CuPC layer, cluster ions should be used to profile this material. Such a profile, using argon cluster ions, is shown in Figure 7.34a. The C : N ratio through the layer is constant, except for the region near the surface where the ratio is slightly higher than expected, this is due to surface contamination with adventitious carbon. The constant ratio suggests that the material is not becoming damaged during the sputter profile measurement. Further confirmation of this can be seen in Figure 7.34b which shows a montage of the copper spectra acquired during the profile. At the surface, some reduction of the copper may be seen but this disappears as the profile proceeds. The reduced copper is observed before the sputtering begins and is removed by the ion beam and so it can be said that there is no evidence for ion-induced damage to the material. The behaviour of the Cu(II) satellite is further confirmation that the copper is not being reduced by the ion beam. Additional confirmation that the CuPC is being removed without changing the chemistry of the material can be seen in Figure 7.34c. This shows the C 1s spectrum acquired from the layer after 8 nm of the material had been removed. Analysis of this spectrum shows that it has the expected peak intensity ratios for the benzene and pyrrole components and a full quantification also shows that the material has the expected concentrations of carbon, nitrogen, and copper.

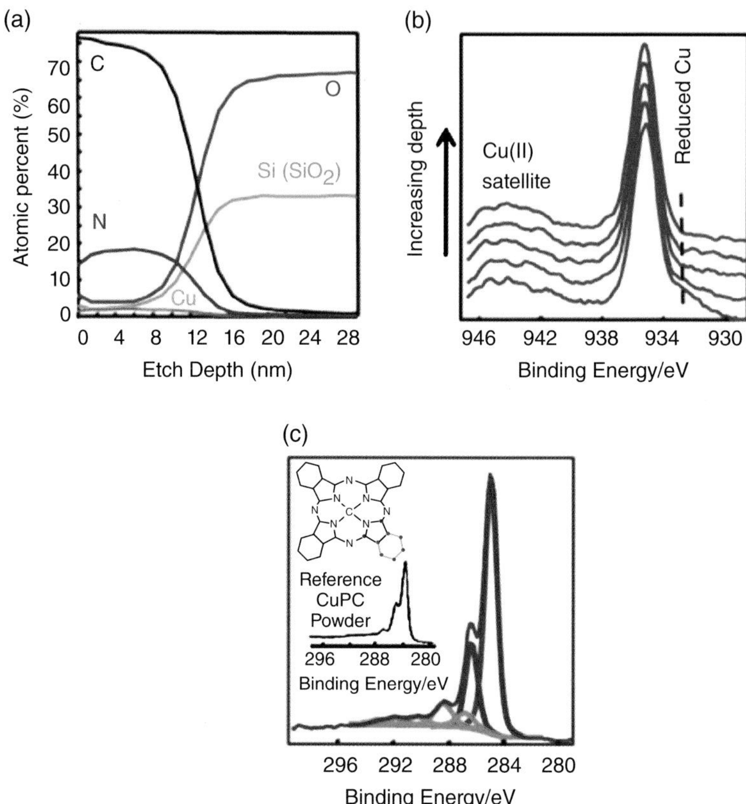


Figure 7.34 (a) XPS profile through the organic field-effect transistor (OFET) using an argon cluster ion beam. (b) A montage of Cu 2p spectra acquired during the profile. (c) A C 1s spectrum acquired after the removal of ~8 nm of CuPC. A reference spectrum of copper (II) phthalocyanine (CuPC) is shown as an inset along with a structural formula of the material.

For a complete analysis of the structure, the depth profile should include the $SiO_2$ layer and its interface with the Si substrate. The $SiO_2$ layer is relatively thick (120 nm) and sputtering is slow using argon cluster ions. It is better, therefore, to sputter this layer using monatomic argon ions. If an ion gun similar to the one illustrated in Figure 4.28 is used, then it is a simple matter to switch from cluster to monatomic ions during the profile. The result of doing this is shown in Figure 7.35.

This clearly shows that it is possible to get an accurate XPS analysis of this layered structure and its interfaces using a combination of cluster ion sputtering and sputtering with monatomic argon ions.

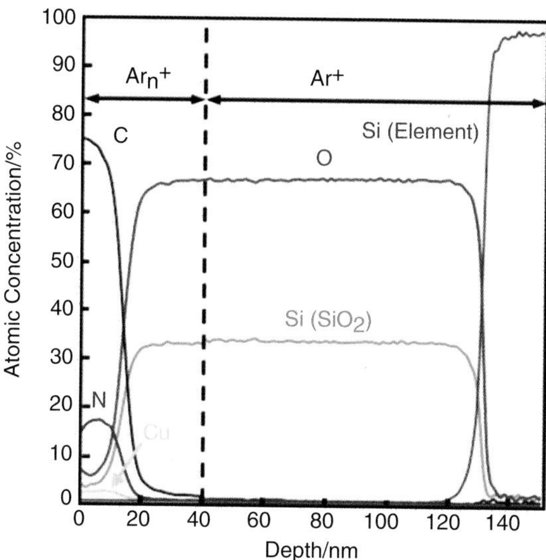


Figure 7.35 A complete depth profile through the organic field-effect transistor (OFET) device to the silicon substrate. The first 40 nm was profiled using an argon cluster beam while the remainder was profiled using monatomic argon ions.

## 7.6 Polymeric Materials

Since its inception as a commercial technique fifty years ago, XPS has been used widely as a method for the surface chemical analysis of polymers. The XPS chemical shift has been shown to be a versatile parameter to describe surface effects in such materials. These may be the result of segregation of minor components or co-polymer constituents, modifications that may occur on surface treatment, such as those used to enhance adhesion and the orientation of molecules at surface. ARXPS is particularly useful in such investigations as changes may only occur in the outer few nanometres of the polymer; a depth scale which matches that achievable with ARXPS almost exactly. However, all such endeavour relies on the existence of a number of databases of polymer XPS spectra recorded at high spectral resolution. It is such data that form the cornerstone of the use of XPS for the characterisation of polymeric materials.

As all organic polymers contain substantial quantities of carbon, it is the chemical shift of the carbon 1s electrons which predominates the interpretation of XPS data from these materials. Figure 7.36 shows the C 1s spectrum of poly(methy methcrylate) recorded using a monochromatic Al K$\alpha$ source. In order to achieve satisfactory peak synthesis of the experimental spectrum it is necessary to use four singlets. These peaks correspond to aliphatic carbon at a

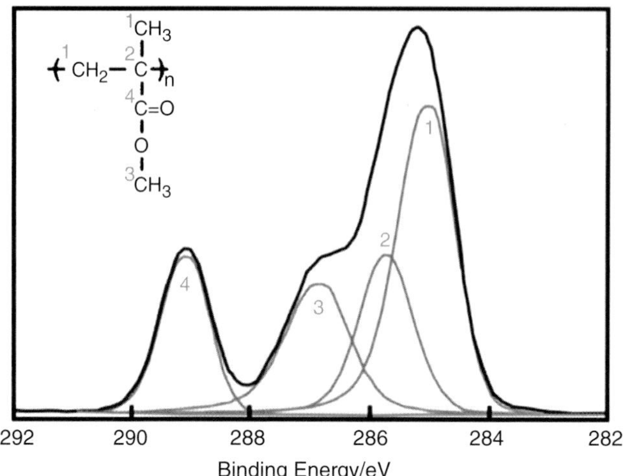


Figure 7.36 C 1s spectrum of poly(methyl methacrylate) recorded with a monochromatic AlKα source. The structural formula of this polymer is also shown and the carbon atoms identified to indicate their contribution to the spectra.

binding energy of 285 eV (a useful internal standard in polymer analysis, but please see the words of caution in Section 3.3.1), carboxyl carbon at a separation of approximately 4.2 eV from the C–C/C–H peak, and ether-like carbon at a distance of about 1.8 eV (as indicated on the structural formula of this polymer, which is also included in Figure 7.36).

The final component at a separation of 0.7–0.8 eV is due to a secondary chemical shift, which is the effect of the carboxyl group on the unsubstituted carbon atom in the C-$CO_2$R structure. The consideration of such secondary shifts (also known as nearest neighbour effects) results from improvements in peak fitting methods and the widespread use of monochromatic X-ray sources. Peak fitting is invariably carried out by computer methods but much of the early work was achieved using much less sophisticated techniques. In such instances, the secondary shift is accounted for by merely making the methyl carbon peak slightly wider and more intense. Unless the carboxyl peak is fairly strong, the secondary shift is easily lost in the vagaries of the peak fitting exercise when using non-monochromated radiation, although the situation is much more secure in the higher resolution spectrum, obtained with a monochromatic source.

By careful use of XPS, it is possible to differentiate between aliphatic and aromatic carbons, there are two possible ways in which this can be done. In the case of high-resolution XPS a small, but significant, negative chemical shift of about 0.4 eV occurs in the aromatic species relative to aliphatic unfunctionalised carbon atoms. The spectrum of Figure 7.37 is taken from polystyrene and the shakeup satellite resulting from the $\pi \rightarrow \pi^*$ transition in the phenyl ring, which

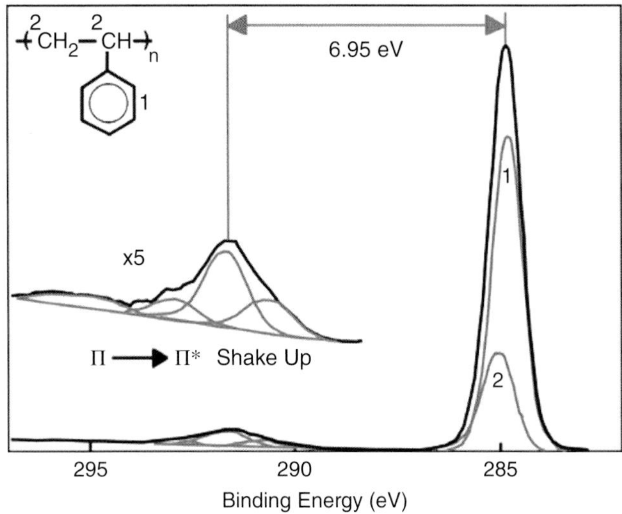


Figure 7.37 C 1s spectrum of polystyrene showing the $\pi \rightarrow \pi^*$ shakeup satellite.

accompanies photoemission, can be seen as a discrete feature some 6.7 eV from the main peak. The intensity of the satellite as a function of the main photoelectron peak remains constant at about 10% although slight changes occur depending on the structure of the polymer involved. This feature provides a quantitative way in which the surface concentration of phenyl groups, following a particular treatment method, may be estimated. It also provides a means of estimating surface modification brought about by ring opening reactions.

In addition to the use of C 1s and other core levels, the VB of the XPS spectrum can be particularly useful for polymers. Once again, the monochromatic source is recommended but this region can often provide very useful, qualitative, information. The XPS VB spectra of poly(ethylene) and poly(propylene) are illustrated in Figure 7.38 and although the C 2p regions are very similar clear differences in the C 2s region of the spectrum are seen. For poly(ethylene) the region 12–25 eV is composed of two readily identifiable features, whilst for poly(propylene) there are three clear components in this region of the spectrum. The C 1s core levels are very similar although both show slight broadening as a result of vibrational effects.

In many instances, VB XPS spectra of polymers are used in a 'fingerprinting' manner to differentiate between similar systems but it is also possible to compare the experimental spectra with calculations made using cluster calculations or other numerical approaches. For practical surface analysis, the former comparison is often sufficient, as one should not underestimate the complexity or time-consuming nature of adopting the second approach!

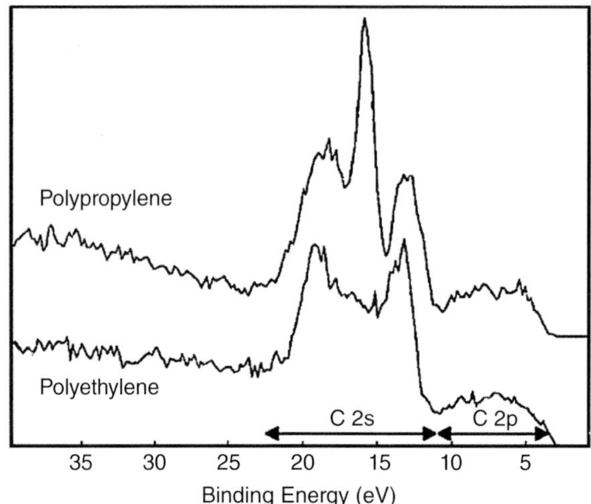


Figure 7.38 Valence band (VB) regions of polyethylene and polypropylene.

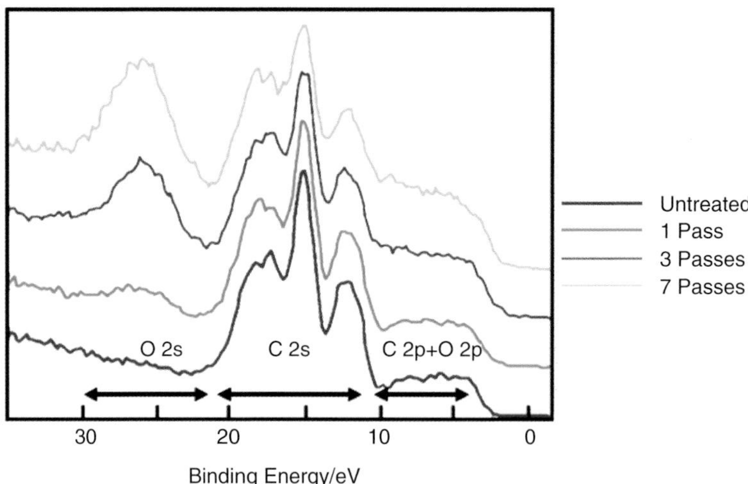


Figure 7.39 XPS valance band (VB) of poly(propylene) following increasing levels of flame treatment. (*Source:* reproduced from Williams et al. [2015], with permission).

An interesting example of the use of the XPS valance band in polymer analysis is provided by work which was carried out to understand the effects of flame treatment on an automotive poly(propylene) material. Flame treatment is carried out to increase the wettability (surface free energy) of the material in order to enhance the adhesion of paints of adhesives. Figure 7.39 shows the VB of this material as a function of the extent of flame treatment. The lowest spectrum is the untreated material, the level of treatment increasing upwards through the montage.

The VB of the untreated material is extremely similar to that of the poly(propylene) standard of Figure 7.38. As flame treatment proceeds so the feature at ca. 15 eV, representative of the pendant methyl group, reduces in intensity. Such an observation is consistent with the removal of the methyl group on flame treatment, as one might expect. The O 2s at a binding energy of 26 eV increases gradually with the level of surface treatment indicating an increasing level of oxygen incorporation. Such oxygenation leads to the gradual increase in polar groups at the polymer surface (readily identified in the C 1s spectrum) and a concomitant increase in surface free energy, responsible for the improved adhesion.

Although XPS has been widely used to study thermoplastic polymers such as poly(methyl methacrylate) and poly(styrene) its use to study crosslinked systems is not so widespread. Such materials are widely used as organic coatings and adhesives which are invariably sophisticated formulations of many components to provide the required mechanical, thermal, process and aesthetic properties. The many organic components ensure that the resultant C 1s spectrum is extremely complex and in order to resolve all the components in the formulation it is necessary to carry out XPS at the highest possible resolution. The following example indicates the complex mature of C 1s spectra from even very straightforward organic coatings. A thermally cured urea formaldehyde/epoxy coating, containing the components indicated in Figure 7.40, was analysed by XPS using monochromatic Al Kα radiation.

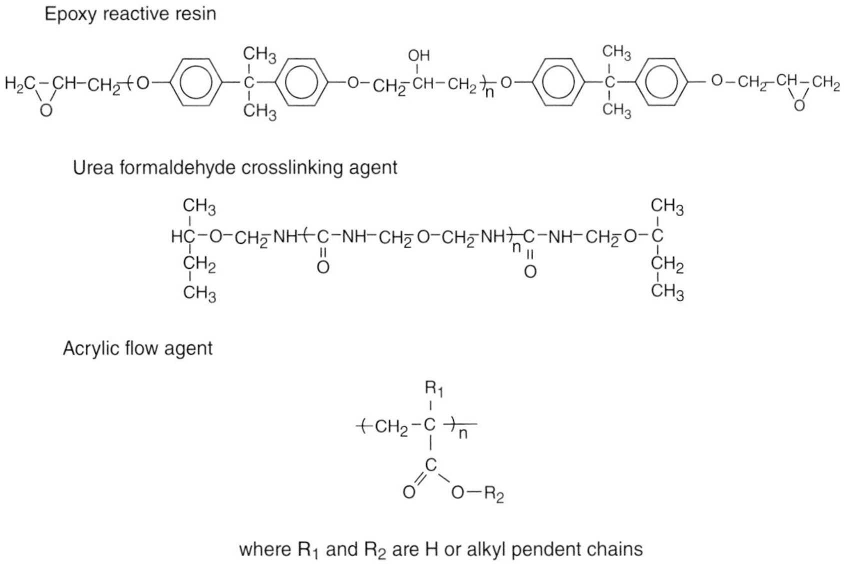

Epoxy reactive resin

Urea formaldehyde crosslinking agent

Acrylic flow agent

where $R_1$ and $R_2$ are H or alkyl pendent chains

Figure 7.40 Components of a thermally cured urea formaldehyde/epoxy coating.

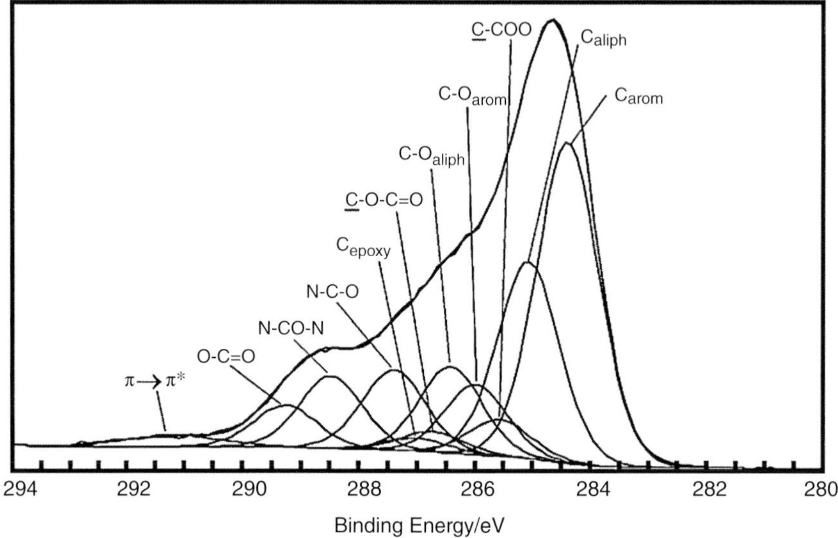


Figure 7.41 XPS spectrum from a thermally cured urea formaldehyde/epoxy coating. (*Source:* reproduced from Perruchot et al. [2002], with permission).

Formulations with and without an acrylic flow agent were examined. The resultant spectrum of the formulation containing the flow agent can be peak fitted to identify the presence of 11 separate components, Figure 7.40. Eight of the components of arise from the urea formaldehyde/epoxy coating, but three uniquely identify the flow agent. These peaks are assigned to carbon atoms in β-position from a carboxyl groups ($\underline{C}$—COO, 285.49 eV), an ester carbon component ($\underline{C}$—O—C=O, 286.67 eV) and carbon species involved in carboxyl groups (O=$\underline{C}$—O, 289.16 eV) respectively.

To investigate the near surface depth distribution of the various elements and carbon functionalities ARXPS was carried out on coating with and without the flow agent.

Such an angle resolved data set does not, in itself, yield a compositional depth profile and a suitable algorithm is required to reconstruct the depth profile from quantitative XPS results. One such method of achieving this end is the routine ARCtick available from the NPL in the UK (www.npl.co.uk). This routine has been used on an ARXPS data set from the urea formaldehyde/epoxy plus flow agent coating discussed above, and the resulting depth profile is shown in Figure 7.42. As expected, the nitrogen concentration, indicative of the coating itself, is depressed to a depth of about 1 nm below the surface as a result of the surface segregation of the acrylic flow aid, which forms a layer about 1 nm thick at the surface.

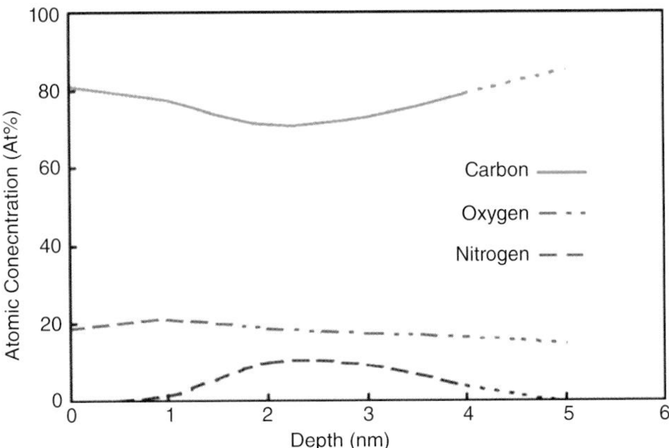


Figure 7.42 Reconstructed depth profile from angle resolved XPS (ARXPS) data of a urea formaldehyde/epoxy coating showing the surface segregation of the acrylic component. (*Source:* reproduced from Perruchot et al. [2002], with permission).

The alternative method of depth profiling, changing the energy of the X-ray photons is potentially a very elegant way with which to probe different depths. In practice, however, it is rather restrictive and most spectroscopists will be limited to a combination of Al K$\alpha$ and Mg K$\alpha$ which gives a depth selectivity of about 1 nm on the C 1s line. As the photon energy increases so the cross section for the lightest elements (including carbon) fall away and although Ag L$\alpha$ radiation can be used for such depth profiling, higher energy sources (e.g. Cr K$\alpha$) are not always successful for depth profiling polymers and the significantly more attractive option is depth profiling using a gas cluster ion source.

With the greater availability of gas cluster ion sources (usually optimised for argon clusters) depth profiling of polymer materials by ion sputtering is becoming much more widely used, and depths approaching one micrometre are now achievable with this method. An important step has been the standardisation of such a method and Figure 7.43 shows the type of sample that has been used for this purpose. This multilayer organic material comprises of a series of layers, of precisely known thicknesses, of two organic materials comprising of H, C, and O (Irganox 1010) and either another material containing these elements and F ($F_{moc}$PFLP[3]) or N (Irganox 1098). In some layers, the two materials are co-deposited and thus the characteristic element (F or N) is reduced in the XPS depth profile.

XPS sputter depth profiles (using 5 kV $Ar_{1000}$ clusters) for the diagnostic elements (N-Irganox 1098 and F-$F_{moc}$PFLP) of the two standard materials is shown in Figure 7.44. As the F and N containing layer structures are nominally the same the N and F depth profiles are very similar in form although the

3 Fmoc-pentafluoro-L-phenylalanine.

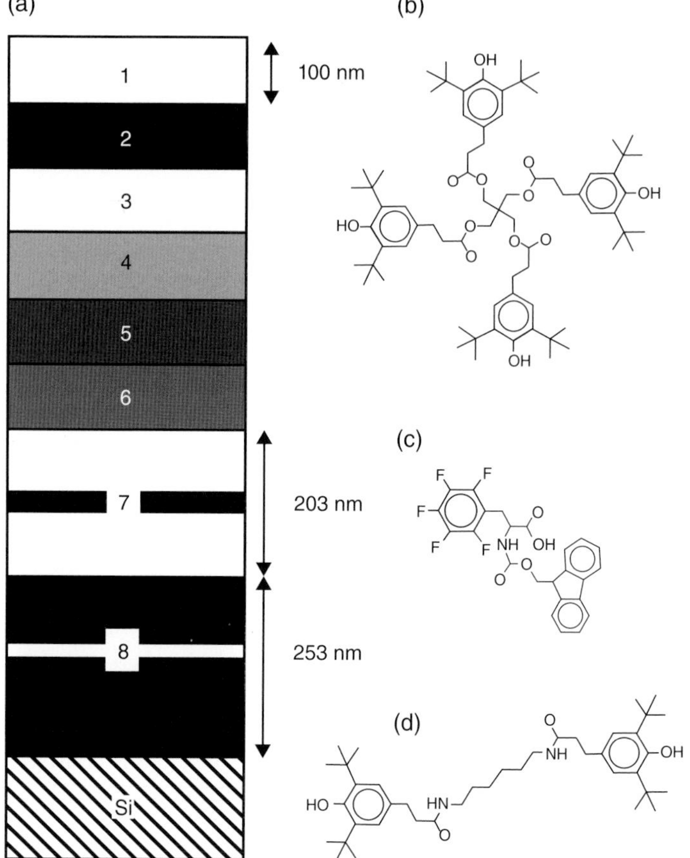


Figure 7.43 (a) Schematic of multilayer sample for sputter profiling of polymers. White is the H, C, and O material, black the one with either F or N as a marker, grey is a mixed layer. (b) Structure of the H, C, O material (Irganox 1010), (c) structure of material with F ($F_{moc}$PFLP), (d) structure of material with N (Irganox 1098). (*Source:* reproduced from Shard et al. [2015], with permission).

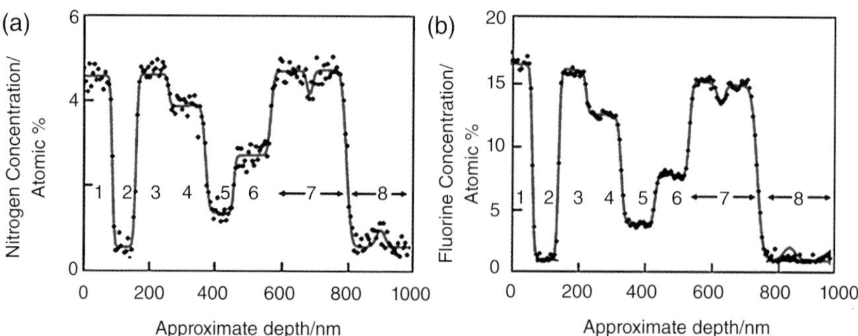


Figure 7.44 XPS sputter depth profiles with argon cluster ions of samples of the type illustrated in Figure 7.43a sample with N containing interlayer, (b) sample with F containing interlayer. (*Source:* reproduced from Shard et al. [2015], with permission).

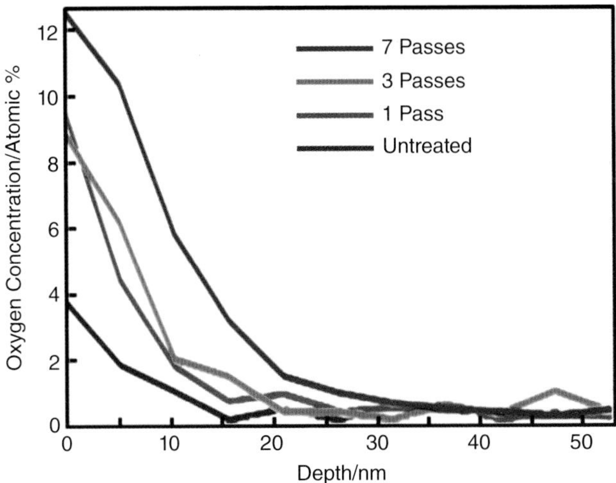


Figure 7.45 XPS Depth profiles of automotive poly(propylene) as a function of extent of flame treatment. (*Source:* reproduced from Williams et al. [2015], with permission).

concentrations of the two characteristic elements are different, reflecting the amount of each element in the respective molecules.

Although the data of these standard calibration samples provide a well-defined multi-layered sample this is often not the case in polymer research, a gradual concentration gradient being more usual. Returning to the example of the automotive poly(propylene) sample discussed above (Figure 7.39), the extent of the oxygen incorporation into the bulk can be determined by depth profiling with argon cluster ions. Figure 7.45 shows sputter depth profiles (using 6 kV $Ar_{1000}^+$ clusters) for the untreated polymer and for increasing levels of flame treatment.

The surface concentration of oxygen increases as a function of flame treatment but, perhaps more importantly, so does the depth to which treatment is achieved. Flame treatment is known to be a relatively long-lived polymer surface treatment compared with, for instance, corona discharge treatment; the reason for this is now known to be associated with the greater depth of treatment of the former treatment.

The vast majority of polymer science investigations making use of XPS report the changes that have occurred as a result of surface modification, either by process treatment (to improve surface properties such as wettability), or naturally occurring phenomena (such as the weathering of paint films). Recent work, however, has concentrated on the identification of surface segregation and depletion phenomena in investigations involving multi-component systems.

## 7.7 Adhesion Science

There are three distinct areas in which surface analysis has made major contributions in the science and technology of adhesion. The analysis of surfaces prior to application of the coating or adhesive, and the subsequent correlation of adhesion with surface cleanliness; the investigation of the substrate to polymer bond; and the exact definition of the locus of failure following bond failure. Each of these areas will now be considered in turn.

The cleanliness of a metal substrate prior to its contact with a polymer adhesive or coating, is readily assessed by XPS or AES. Although gross levels of contamination such as oils from mechanical working or temporary corrosion protection can easily be detected by other methods, it is only surface sensitive techniques which can assess the efficacy of a cleaning method. As an example, Figure 7.46 shows spectra taken from steel sheet that has been prepared by alkaline cleaning (Figure 7.46a) and emery abrasion (Figure 7.46b). The level of carbonaceous contamination is considerably higher on the solution-cleaned surface. This indicates that abrasive cleaning produces the better-quality surface from a chemical point of view. Studies some thirty years ago by the American automobile industry established unequivocally that high levels of

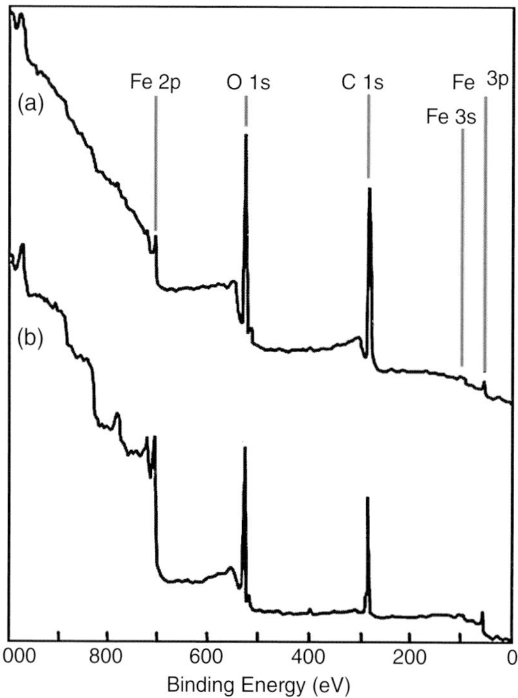


Figure 7.46 Surface cleanliness of steel sheet. An alkali cleaning process (a) leaves substantially more contamination than an abrasive cleaning process (b).

surface carbon are a contributory factor to poor durability of painted steel from certain manufacturers. Another frequently cited example is that of aluminium-magnesium alloys, which, if heat-treated incorrectly, develop a friable surface film of magnesium oxide. Adhesion of a paint film or adhesive to such substrates is very poor and failure occurs rapidly. XPS or AES is able to identify such a layer and these techniques can be used in a diagnostic manner prior to bonding or painting to ensure that adhesion will be of the required standard. Contamination may also arise from a variety of external sources, sub-monolayer coverage of aluminium surfaces by fluoro-carbons has been shown to produce a drastic reduction in adhesive bond strength.

Once the bond has been made, the task of examining the interfacial chemistry is extremely difficult. Various methods of approaching the interface have been developed, involving the removal of the polymer in a suitable solvent, or the dissolution of the iron substrate in a methanolic iodine solution followed by sputter depth profiling through the oxide towards the interface, but both suffer from their own particular problems. Careful ultramicrotomy followed by scanning transmission electron microscopy (STEM) used in conjunction with windowless EDX, electron energy loss spectroscopy (EELS), or electron diffraction may be more useful where an interphase has developed, but the analysis of the interface on an atomic scale by these methods is still some way off. However, the combination of argon cluster profiling and XPS is a much more promising approach for the analysis of substrate/adhesive interface with molecular resolution.

One productive approach to the probing of the interface chemistry directly is the use of thin layers of model compounds deposited from very dilute solutions or even the use of dilute solutions of multi-component commercial products. In this manner, it is possible to determine the orientation of the molecules at the surface using ARXPS and deduce the nature of chemical bonding by interpretation of the C 1s (and others) chemical shift. As an example of the latter method Figure 7.47 shows a set of C 1s spectra recorded from thin films ($<2$ nm) of poly(methyl methacrylate) applied to various oxidised metal substrates.

Subtle differences in the C 1s spectra are observed and are ascribed to the nature of the interactions between the polymer and the oxide substrate. The three substrates are silicon, aluminium, and nickel whose oxides are acidic, weakly basic, and strongly basic respectively. The interactions that occur are shown in Figure 7.48 and are hydrogen bonding (silicon), bidentate interaction (aluminium), and acyl nucleophilic attack (nickel), which are established on the basis of the C 1s spectra of Figure 7.47.

Polyurethanes are important materials for a whole variety of applications ranging from adhesives and coatings through to foams used in furniture and sandwich panels. Polyurethane chemistry is based on the reaction between polyols and isocyanates and in order to understand and engineer the adhesion

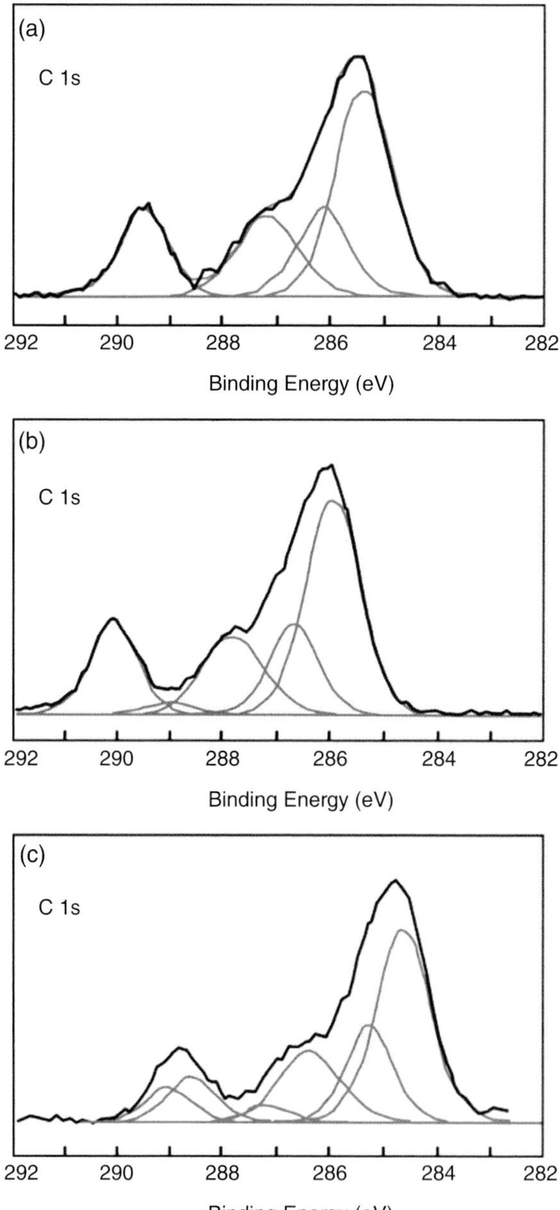


Figure 7.47 C 1s spectra of thin films of PMMA applied to oxidised metal substrates of different acid–base character. (a) silicon, acidic, (b) aluminium, weakly basic, (c) nickel, strongly basic. (*Source:* reproduced from Leadley and Watts [1997], with permission).

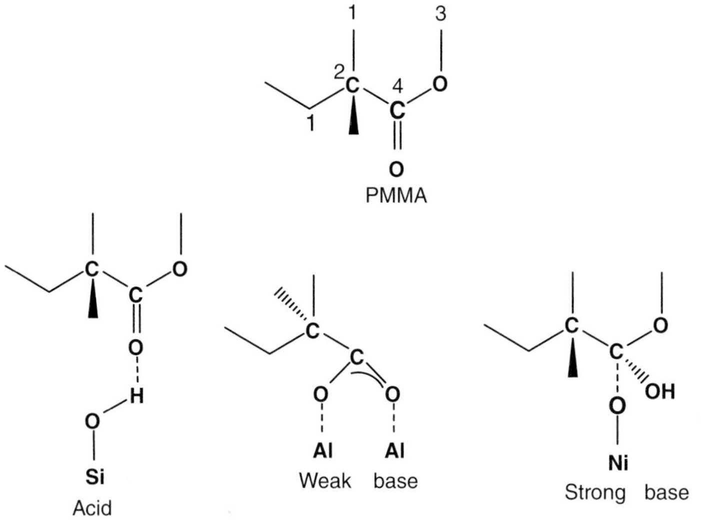

PMMA

Si Acid

Al Al Weak base

Ni Strong base

Figure 7.48 Interactions responsible for the fine structure in the C 1s spectra of Figure 7.47. Numbering of carbon atoms on the upper schematic refer to C 1s components assigned in Figure 7.36.

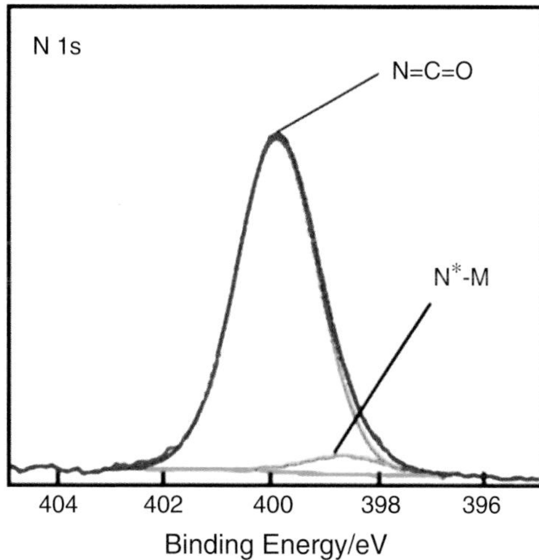


Figure 7.49 N 1s spectrum of poly(methylene diphenyl diisocyanate) (pMDI) adsorbed on a Fe/Cr alloy.

between polyurethanes and metallic substrates it is useful to study the interaction of isocyanates with the surface of technological metals. The following example illustrates this approach in the study of the reaction between poly(methylene diphenyl diisocyanate) (pMDI) and a Fe75Cr25 alloy. In a similar vein to the PMMA example, above, Figure 7.49 shows the N 1s spectrum of a very thin layer of pMDI applied from a 0.05 w/v% solution in acetone on an oxidised Fe75Cr25 substrate. At this concentration the layer deposited, which was cured at elevated temperature after deposition, is thin enough that XPS can be used to study the interface chemistry directly. The N 1s spectrum shows two clear components; the major component at ca. 400.0 eV can be attributed to the isocyanate group in the organic structure, whilst the lower binding energy component (at ca. 398.5 eV) is a result of a specific interaction between the isocyanate molecule and oxidised metal surface identified as N*-M, where M is a metal. The occurrence of such an interaction sees polarisation of the electron cloud towards the nitrogen atoms of the pMDI, bringing about a shift to a lower biding energy, in much the same manner as one would see on the formation of a metallic nitride.

The N*-M component is very small and this treatment may not be sufficiently rigorous to satisfy some scientists. In order to substantiate such results and add a degree of objectivity to the peak fitting, a complementary approach was adopted using a much thicker layer of pMDI deposited from a 1 w/v% solution on the same Fe75Cr25 substrate. In order to access the interfacial chemistry in this case it was necessary to use argon cluster ions to sputter through the pMDI overlayer to access the interface. Once collected, the spectra were processed by a multivariate analysis method called non-negative matrix factorisation (NMF). This enables the chemical information contained in the N 1s spectra to be extracted automatically. The advantage of this approach is that NMF is an unsupervised dataprocessing method and so provides a much greater degree of objectivity to the process of unravelling the chemical state information. NMF identifies three components in the N 1s spectral region (394–406 eV); the pMDI bulk, the interface region, and a background contribution from inelastically scattered electrons. The relative intensities of these three components as a function of argon cluster ion sputter time is presented in Figure 7.50.

The depth profile is comprised of fifty individual analyses, or levels. By considering the spectra generated by NMF at each level, those representative of bulk pMDI, interfacial regions and deep into the profile, where most of the pMDI has been removed, it is possible to establish the XPS signatures for the pMDI and the interface. This is shown in Figure 7.51, and the agreement between the pMDI and interface components of the single N 1s spectrum of Figure 7.49, is readily apparent. This provides added confidence in the fidelity of the peak-fitting process and enables a reaction mechanism of pMDI with Fe75Cr25 to be proposed.

The reaction that occurs at the interface is thought to be a two-stage process. The isocyanate group of the pMDI reacts with water and is reduced to a primary amine with the liberation of carbon dioxide. The amine species then reacts with the (hydroxylated) metal oxide surface to form a N*-M bond, with the liberation of a hydroxonium ion, charge from the cation being drawn

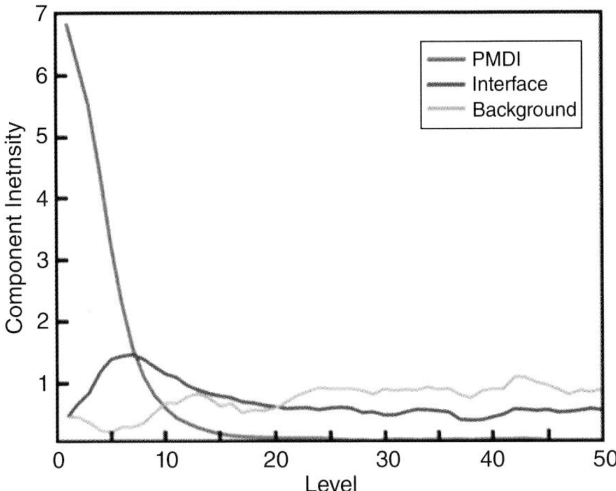


Figure 7.50 Argon cluster ion depth profile of a thick layer of poly(methylene diphenyl diisocyanate) (pMDI) on Fe/Cr alloy. Three components are identified by non-negative matrix factorisation (NMF) representing the bulk pMDI, the interface region and the energy loss background.

towards the nitrogen atom leading to a reduction in binding energy of this species, as shown in Figure 7.52.

Using approaches such as these, it is possible to track the interfacial chemistry that is responsible for the adhesion of polymers to metal substrates. In many cases, the complementary molecular information provided by time-of-flight secondary ion mass spectrometry (ToF-SIMS) is able to add an extra dimension to our understanding of such chemistry.

In this manner, a great deal of information has been obtained in recent years concerning the manner of interaction of organic molecules, relevant to adhesives and organic coatings, with oxidised metal substrates. An important complementary technique in this work has been high resolution ToF-SIMS. Indeed, the combination of XPS with ToF-SIMS is very powerful for the comprehensive definition of polymer surface chemistry and the interaction of organic molecules with solid substrates. XPS can also be used to monitor the capacity of a solid surface for species in the liquid phase by the construction of adsorption isotherms from XPS (or ToF-SIMS) data. This has the advantage that the uptake at a variety of solution compositions can be recorded directly by measurement of the surface composition, the uptake curve is simply the surface composition as a function of the solution composition.

The analysis of a failed interface is routinely carried out by electron spectroscopy and the definition of adhesive or cohesive failure, on a molecular scale, has become a straightforward matter for those working in the field. The

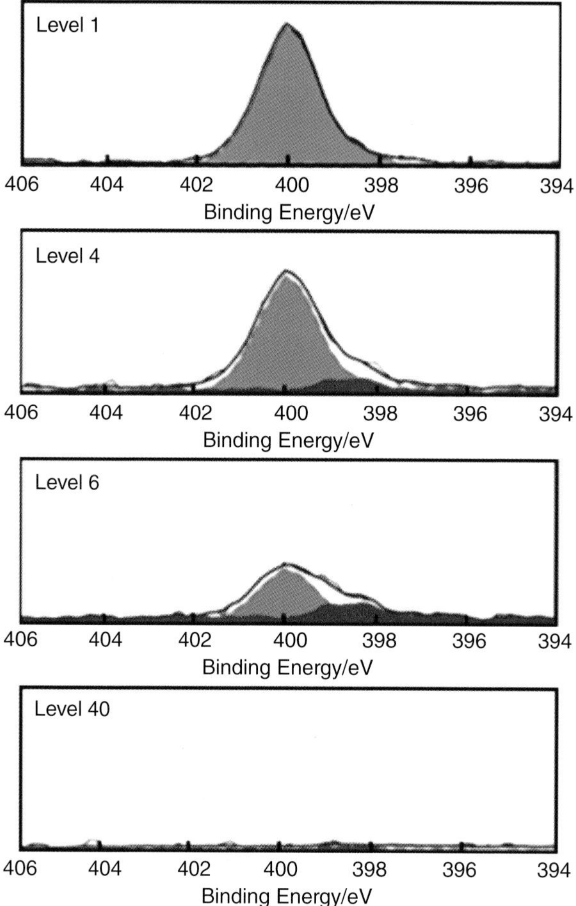


Figure 7.51 Non-negative matrix factorisation (NMF) results for N 1s spectra for levels of the depth profile of Figure 7.50. The green component represents poly(methylene diphenyl diisocyanate) (pMDI) and the red one a component characteristic of the interfacial chemistry that occurs between pMDI and an Fe/Cr alloy.

spectra of Figure 7.53 represent analyses, at a spot size of 400 μm, taken from the failure illustrated in the same figure. The system concerned is a tin-free steel (TFS) substrate coated with an epoxy-phenolic lacquer which in turn has been joined, by a hot melt process, to a nylon substrate. The practical application of this system is the bonding of gas capsules ('Widgets') to the internal surfaces of metal beverage cans. The visual appearance of the complementary mirror images of the failure surface is quite striking. The nylon side of the failure is seen as a white colouration, (RHS) of Figure 7.53 with a dark patch on the lower left, the coated TFS substrate (LHS) of Figure 7.53 is darker with a

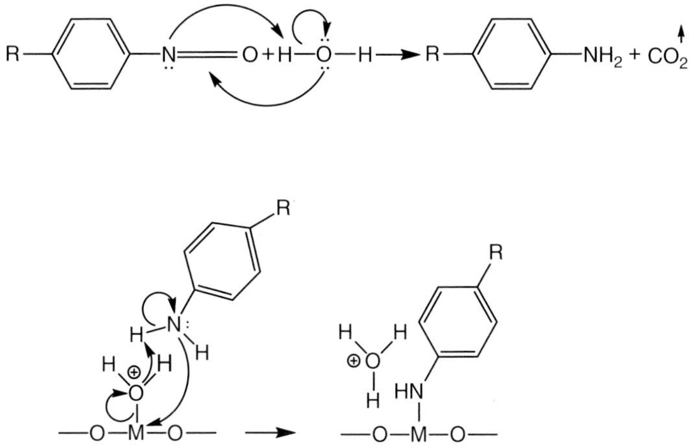

Figure 7.52 Reaction of poly(methylene diphenyl diisocyanate) (pMDI) with a hydroxylated metal surface. Upper scheme: Initial reaction of pMDI with water to yield an amine, Lower scheme: interaction of modified pMDI with metal surface.

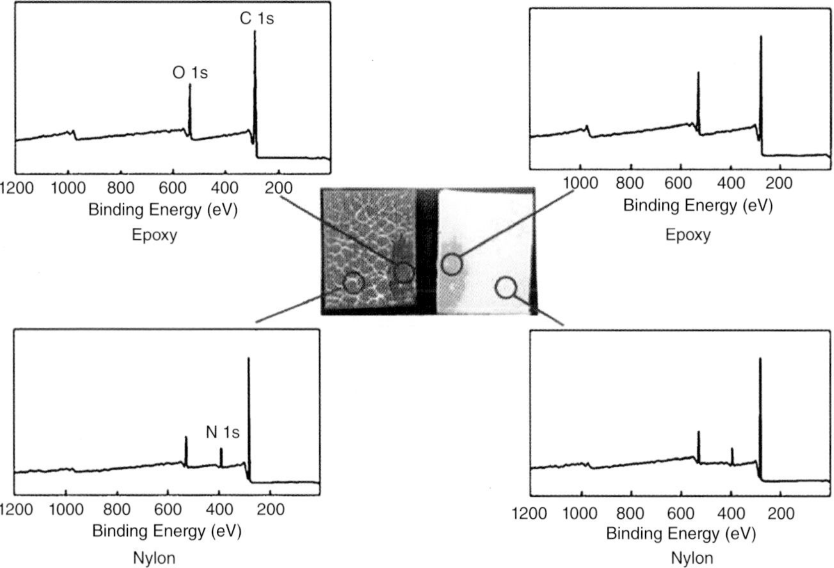


Figure 7.53 Failure of a TFS/epoxy-phenolic/nylon system.

white, lace-like deposit together with a much darker region on the lower right. The spectra recorded from the minority region are characteristic of epoxy and represent cohesive failure within the TFS lacquer. The failure between nylon (white) and epoxy is, however, shown to be cohesive within the thermoplastic, as shown by the similarity of the lower spectra of Figure 7.53. Visual observations alone would have identified these failures as interfacial between the lacquer and TFS, and interfacial between the nylon and the lacquer.

As indicated by the examples chosen above, it is XPS rather than AES that is most widely used in adhesion studies, often in conjunction with ToF-SIMS. Another area of intense activity over the last decade or so has been their use in studies of composite materials. XPS has been used for some time to assess the surface acidity of carbon fibres and the level of sizing. *In-situ* fracture within the spectrometer can now be achieved quite readily on carbon fibre composite materials. Already, both XPS and AES have been used to study the interfacial region in metal matrix composites, and it appears that, in some cases, minor elements from the alloy matrix may segregate to the (ceramic) fibre surface.

## 7.8 Nanotechnology

The approaches that can be taken when dealing with nanotechnology structures depend on whether a 'bottom-up' or a 'top-down' approach is to be taken in the fabrication of the nanostructure. The 'bottom-up' process involves the fabrication from a starting material of very fine particles or other entities which are then consolidated in some manner to produce the required nanostructure with features on the sub-micrometre scale. In the 'top-down' approach a subtractive method is used to remove the material that is not required. This can be considered as engineering on the sub-micrometre scale. As far as the use of electron spectroscopies are concerned, the bottom-up source material is very amenable to analysis by XPS, impurities can be identified at surfaces, oxide layers on metal nanomaterials described in terms of thickness and composition and modifications resulting from environmental exposure followed as a function of time. In nanomaterials a very high proportion of the atoms, ions or molecules making up the material will be at the surface, so a surface-specific technique becomes essential. Any structure–property relationships associated with nanomaterials must necessarily probe the surface composition rather than the entire volume of the particle. Thus, an important role for XPS is assured, notwithstanding its modest spatial resolution, as it is invariably feasible to assemble a raft of particles or a longitudinal arrangement (a tow) of fibres. In the case of top-down nanostructures XPS is of relatively limited use as the size scale will be well below that achievable with XPS and AES becomes the method of choice. One possible route to the application of XPS in such a nanostructure, if there are insulating and conducting regions, is to apply a potential of a few volts between sample and earth which will move the spectrum from the conducting parts of the structure a comparable number of eV up the energy scale. Although this can be a useful approach, the spatial resolution of AES, now reaching values below 10 nm is generally considered a more reliable route to the surface analysis of samples of this type.

Perhaps the most straightforward form of heterogeneous nanoparticle to analyse is that of a sphere covered by a thin layer of a chemically different material, a so-called 'core-shell nanoparticle'. Work by XPS on samples of this form were reported more than four decades ago but, in recent years, the level of precision that is attainable with such measurements has improved significantly as the theoretical treatment has developed. The placement of shell thickness measurements on a sound basis that can be readily adopted by the workaday surface analyst probably started with the definition by NPL in the UK of XPS *Topofactors* used in conjunction with an earlier method to determine the overlayer thickness from planar samples (*The Thickogram*). The *Thickogram* is essentially a nomographical scale to provide a measure of overlayer thickness, t, (in units of $\lambda \cos \theta$: where $\lambda$ is the attenuation length of electrons in the overlayer), obtained using appropriate forms of the Beer Lambert Equation (Section 1.6). This is achieved by construction of a normalised peak area intensity ratio, of overlayer signal to substrate signal, against a curved logarithmic grid of the intensity ratio against the energy ratio of the two photoelectron peaks in question. The overlayer thickness is quite simply the linear intercept, on an exponential curve defining the quantity $t/\lambda \cos \theta$ between the appropriate points on the logarithmic intensity ratio line and the curved logarithmic grid. The *Topofactors* approach was then developed further into a simple approach where the relative XPS intensities characteristic of the core and the overlying shell can be converted into a straightforward measurement of shell thickness. As the compositional gradients within nanostructured materials become more complex so the analysis inevitably becomes more involved and, to a certain extent, the structural results obtained less secure. However, it is clear that XPS has an important role to play in the analysis of such materials. In addition, extreme care must be taken in the preparation of nanostructured materials for surface analysis, a topic which is the subject of a recent ISO standard.[4]

The investigation of carbon surfaces has taken on increasing importance with the introduction to the scientific community of, in rapid succession, buckminsterfullerene ($C_{60}$), carbon nanotubes (CNTs) and graphene. XPS can have a number of roles to play in such investigations as illustrated by the following examples which illustrate the use of XPS for the assessment of purity

4 ISO 20579-4 : 2018 – Importance of sample preparation on reliable surface characterisation of nano-objects.

of CNTs and the use of the C *KLL* D-Parameter (Section 3.3.4) for assessing changes in the surface chemistry of CNTs.

XPS survey spectra of three samples of CNTs are provided in Figure 7.54, the CNTs being as received (Figure 7.54a), the same CNT following carboxylic acid functionalisation (Figure 7.54b) and a sample of a commercial (rather than research grade) CNT that was rejected as it was heavily contaminated with magnesium oxide and other metals (Figure 7.54c).

The quantitative surface analyses from these three CNT samples are provided in Table 7.4.

As can be seen from the data of Table 7.4, the level of incorporation of carboxylic acid groups is fairly modest. It is at a concentration that is reminiscent of the extent of acidification introduced to carbon fibres by commercial oxidation treatments which is necessary to achieve the required level of adhesion between fibre and matrix in a carbon fibre reinforced polymer composite material. The fibre of Figure 7.54c is a different matter, however, being heavily contaminated with magnesium oxide. The source is unclear but a straightforward XPS analysis was sufficient to persuade the supplier to refund the purchase price!

Graphene is essentially single or multilayer arrays of graphite which has the potential for applications in many industries. After its initial isolation by mechanical cleavage, the challenge arose to produce graphene by more cost-effective production routes. One such route is the treatment of graphite with a strong oxidising agent followed by reduction to graphene. An example of this process is the production of graphene oxide by a modified Hummer process, followed by reduction at a range of conditions to reduce the concentration of oxygen and the direct production of graphene. XPS provides a facile method for the assessment of the efficacy of the reduction process of the flake-like graphene oxide as shown by the survey spectra of Figure 7.55. The original graphene oxide produced by oxidation and exfoliation of graphite is identified as GO and the reduction processes are identified as G1, G2, and G3, increasing numbers being indicative of an increasing level of reduction. The reduction in oxygen concentration is clearly seen in the tabular quantitative data of the inset of Figure 7.55, with 26% on the graphene oxide being reduced to <2% on the example which has been reduced to graphene (G3).

The high-resolution spectra confirm the oxygen functionality and enable the identification of $sp^2$ and $sp^3$ bonding (Figure 7.56a), the former becoming totally dominant as the reduction process becomes more aggressive. A similar trait is observed in the C *KLL* data of Figure 7.56b where the D-Parameter is seen to increase from 17.5 eV, close to diamond, for GO to 20.3 eV, just below graphite, for G3. Using these data, it is a straightforward process to estimate the amount of $sp^2$ character using the linear interpolation approach of Section 3.3.4.

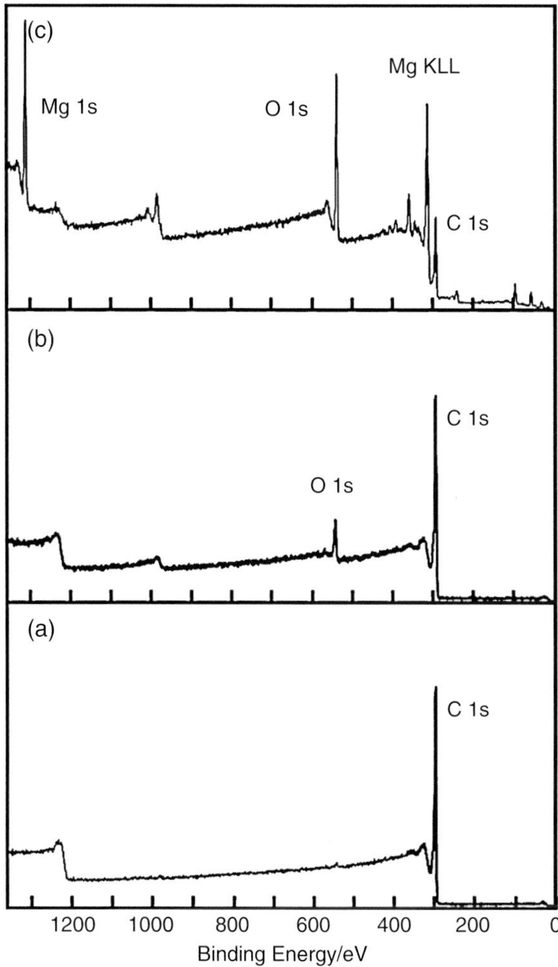


Figure 7.54 XPS survey spectra of carbon nanotubes (CNTs) (a) as received, (b) carboxylic acid terminated and (c) heavily contaminated with magnesium oxide.

Table 7.4 Quantitative surface analysis of carbon nanotubes (CNTs) of Figure 7.54.

| | Surface composition/atomic % | | | | | | |
|---|---|---|---|---|---|---|---|
| | **C** | **O** | **Si** | **Cl** | **Mg** | **S** | **Mo** |
| As received | 99.0 | 1.0 | – | – | – | – | – |
| Carboxylic acid | 90.8 | 8.2 | 0.7 | 0.2 | – | – | – |
| MgO contaminated | 30.5 | 38.5 | 0 | 0 | 29.6 | 0.7 | 0.7 |

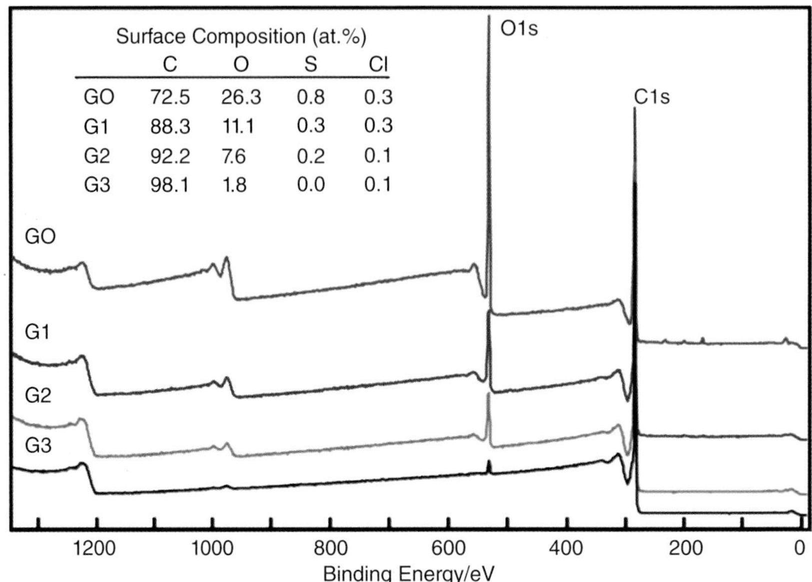


Figure 7.55 XPS survey spectra of graphene oxide (GO) and reduced GO (G1–G3). (This work was conducted at the Center for Nanophase Materials Sciences, Oak Ridge National Laboratory [ORNL], which is sponsored by the Scientific User Facilities Division, Office of Basic Energy Sciences, U.S. Department of Energy).

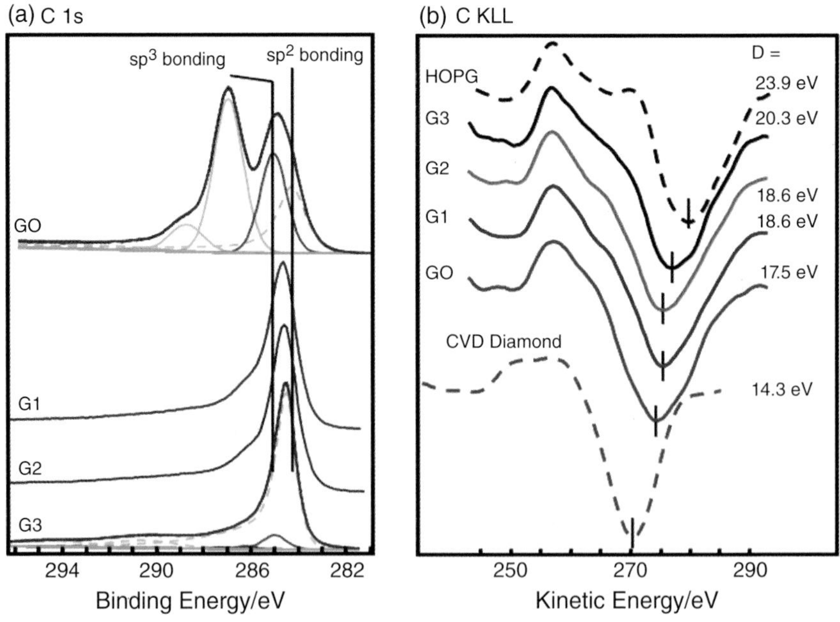


Figure 7.56 Data for graphene oxide (GO) and reduced graphene oxide (G1–G3). (a) C 1s spectra, (b) C *KLL* spectra including highly oriented pyrolytic graphite (HOPG) and chemical vapour deposition (CVD) diamond, including D-Parameter. (*Source:* this work was conducted at the Center for Nanophase Materials Sciences, Oak Ridge National Laboratory [ORNL], which is sponsored by the Scientific User Facilities Division, Office of Basic Energy Sciences, U. S. Department of Energy).

## 7.9 Biology

The application of XPS to biological samples and biomaterials investigations has been an extremely fruitful area of endeavour going back to the very early days following the commercialisation of the technique. Such investigations are, however, rarely straightforward as biological samples in an *in vivo* or *in vitro* condition are invariably associated with large quantities of water. Hardly conducive for introducing into the UHV chamber of a typical XPS system! The approach, until fairly recently, was to either immobilise or remove the water; immobilisation by freezing the specimen, removal by careful freeze-drying or spray drying. The choice of method depends significantly on the nature of the investigation. For instance, there has been a great deal of work on the adsorption of a variety of proteins on implant materials for the human body. The driving force here being that the first physiological medium that an implant (metal, polymer or ceramic) will encounter will be protein rich, thus protein adsorption is the pre-cursor to more important events such as osseointegration. The study of the adsorption of monolayer quantities of a protein on a solid surface is straightforward and does not need any special preparative measures. Larger quantities of such material and synthetic (e.g. hydrogels) and natural (e.g. cells) materials containing large amounts of water must be treated with caution. A standard method is to introduce the sample into the spectrometer in the frozen state and conduct the analysis at cryo-temperatures using liquid nitrogen cooling. Such an approach is not straightforward as there will be a tendency for the sample surface to be covered with a thin layer of ice if chilled in ambient atmosphere and the cooled sample in the spectrometer to act as a site for the preferential adsorption of any hydrocarbon present in the system as a result of its lower temperature. The solution to both these problems is fairly straightforward in that the sample must be chilled in an environment free from water vapour, such as a glove box attached to the XPS system. In the case of contamination within the spectrometer one must ensure a good base vacuum level, an additional cold finger chilled with liquid nitrogen positioned close to the specimen will also do much to alleviate the problem.

The quantitative nature of XPS is particularly useful for the estimation of coverage of a substrate by a protein layer but the identification of specific proteins is not possible by XPS and the user will generally resort to the use of ToF-SIMS in conjunction with multivariate analysis. The determination of the integral quantity of protein by XPS can be deduced from the surface concentration of nitrogen but identification of the distribution, whether island-like or uniform, requires ARXPS. A family of curves is plotted for different surface coverages and a thickness/coverage relationship deduced. A potentially more elegant solution is to combine XPS with atomic force microscope (AFM), the former providing the quantity of protein and the latter the topography. The data of Figure 7.57, recorded at ambient temperature, shows the gradual attenuation of the Ti2p spectrum as a result of exposure of a series of commercially pure (CP) titanium coupons to increasingly concentrated solutions of the protein albumin, as can be seen the solution of $1 \times 10^{-3}$ v/v% concentration is thick enough to totally attenuate the Ti 2p spectrum.

The data of Figure 7.57 is part of a series required to construct an XPS adsorption isotherm describing the interaction of albumin with titanium, showing that monolayer coverage is achieved at a solution concentration of approximately $10^{-4}$ v/v%. Complementary AFM data shows that at this point the protein starts to take on its characteristic globular shape and a gradual increase in surface roughness ($R_a$) is observed from 1.5 nm at a solution concentration of $10^{-4}$ v/v% to 2.3 nm at $10^{-3}$ v/v% and finally to 3.0 nm at concentrations of $10^{-2}$ v/v% and greater.

Spray drying is a process which involves the dehydration of liquid droplets in a stream of hot air and is widely used in the preparation of particulate foodstuffs such as dairy products. For this reason, it has become a popular preparation method with those involved in the XPS analysis of food products, although freeze drying is also popular. Figure 7.58 shows the O 1s, N 1s, and C 1s spectra recorded from model cake formulations (absence of oil or egg products) following freeze drying.

The assignment of the different functionalities identified in the three spectra of Figure 7.58 is challenging and relies on careful comparison with the spectra of the individual components themselves, along with standard compounds, in order to unravel the fate of phospholipids, carbohydrates and egg

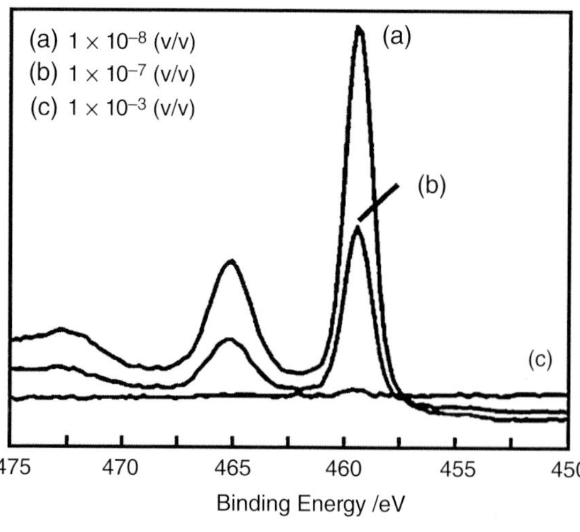


Figure 7.57 Ti 2p spectra of CP titanium after albumin adsorption from the indicated solution concentrations. (*Source:* reproduced from Aeimbhu et al. [2005], with permission).

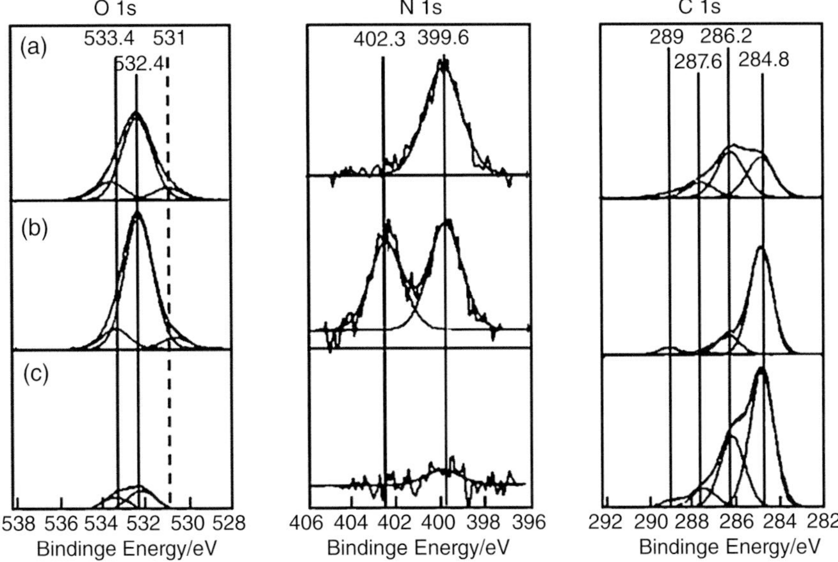


Figure 7.58 O 1s, N 1s, and C 1s spectra of cake formulations of flour and sucrose plus (a) egg white, (b) egg yolk and (c) palm oil. (*Source:* reproduced from Rouxhet and Genet [2011], with permission).

and flour proteins in the formulations. With care this is possible and important information regarding surface characteristics of the individual cake mixes is obtained.

As indicated freezing or dehydration of specimens with large amounts of water is a popular approach to sample preparation for the XPS analysis of such samples, but recent innovations in near ambient pressure XPS (NAPXPS) have made analysis of such samples a more straightforward and, indeed a more certain activity. As an example of such analyses, Figure 7.59 presents the XPS survey spectrum of sample of hard cheese acquired using a commercial NAPXPS system at a pressure of 1 mbar.

This type of sample would be particularly difficult in a UHV XPS system and the operator would need to resort to using a cryostage, with the attendant problems of oils and so forth within the cheese migrating to the sample surface during the freezing process. As an indication of the quality of data that can be obtained with such a system the high-resolution C 1s spectrum is presented below (Figure 7.60).

Interpretation of the fine structure present in the high-resolution spectra allows the presence of the main constituents of the cheese, fat, and carbohydrates, to be identified. The inset of Figure 7.60 shows the generic structure of a triglyceride molecule, a typical fat.

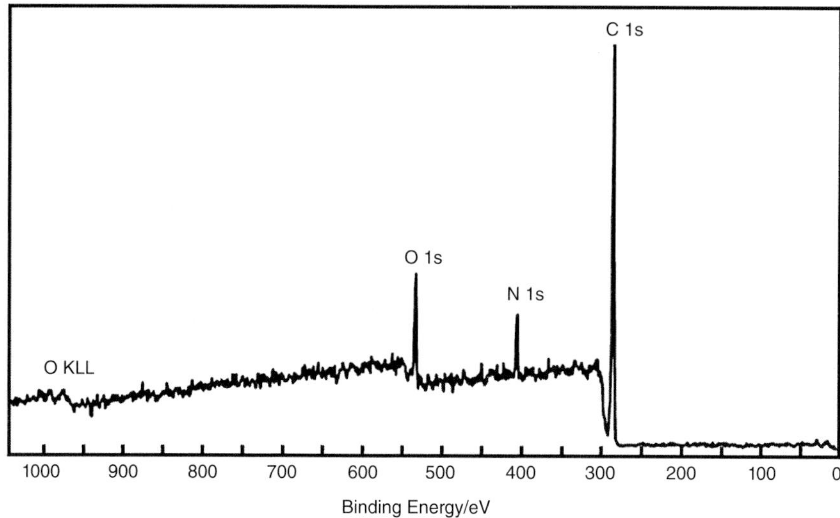


Figure 7.59 XPS survey spectrum of a sample of hard cheese. The spectrum was collected at a pressure of 1 mbar.

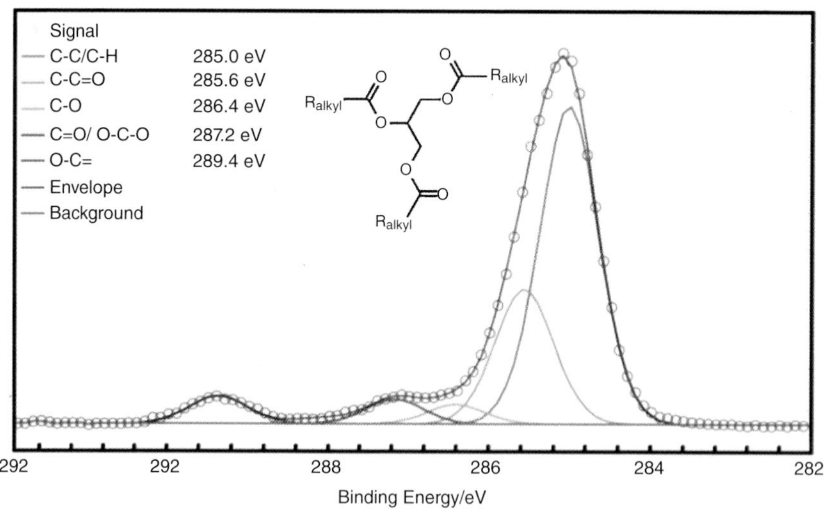


Figure 7.60 C 1s spectrum of sample of hard cheese acquired using near ambient pressure XPS (NAPXPS).

## 7.10 Energy

The electron spectroscopies provide valuable tools for the understanding of batteries, fuel cells and solar cells. Areas where these analytical methods contribute to an understanding of the technologies include:

- Measurement of the thickness and uniformity of layers in multi-layered structures.
- Accumulation of impurities and growth of undesirable films at electrode surfaces.
- Unintentional migration of species from one electrode to the other.
- The nature of the electrode/electrolyte interface.
- Chemical changes to electrodes brought about by repeated charging and discharging of batteries.
- Processes that cause degradation during use and affect the life time of the device.
- Measurement of the work function of solar cell surfaces, checking for uniformity and damage.
- The effect of thermal cycling and annealing upon the components of solid oxide fuel cells.

XPS and Auger have made a major contribution to the understanding of lithium ion batteries. Working with the components of these batteries presents particular difficulties because exposure to the air seriously and rapidly affects the chemistry of the electrode surfaces. For this reason, it is essential to use a glove box and/or a vacuum transfer vessel when handling the samples.

In common with other areas of technology, analysis using depth profiling techniques is very common here and the methods used are very similar. The following example concerns an organic photocell. It is of interest because it requires a depth profile through both an organic layer and an inorganic layer ($SiO_2$) and because, being a photo cell, the valence band is of interest in addition to the core level transitions.

The necessity to maintain the chemical integrity of the organic layer (so that its VB can be studied as a function of depth) means that argon cluster ions must be used to profile the material. Ultra-violet photoelectron spectroscopy (UPS) is superior to XPS for the study of the VB and so both an XPS and a UPS profile is required. The ideal way to do this is to use a multi-technique instrument which has full digital control and to acquire both XPS and UPS data following each etch cycle. The method is indicated in the flow chart, Figure 7.61, for a profile experiment which includes *n* etch cycles. By performing the analysis this way, etching only needs to be done once, saving time, and the analyst can be certain that the XPS and the UPS measurements are taken from exactly the same depth and exactly the same position on the sample. The UV lamp and the

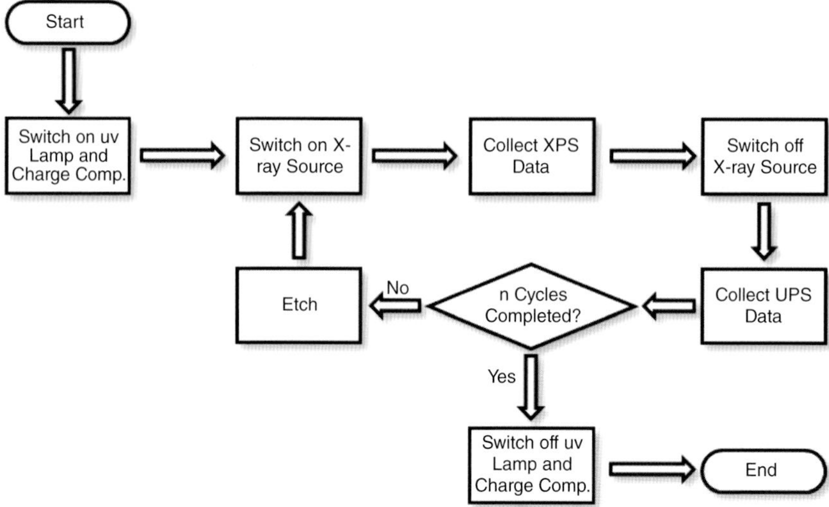


Figure 7.61 A flow chart showing the experimental steps required to produce an XPS and UPS depth profile simultaneously.

charge compensation gun may be switched on at the start of the experiment and not switched off until the end because they will not interfere with any part of the analysis. The analysis can only be done in this way if the instrument has a sufficiently high pumping rate because the ion gun, the charge neutralising gun and the UV lamp all produce a high gas load.

In this example, the organic layer consists of a mixture of phenyl-C61-butyric acid methyl ester (PCBM) and poly(3-hexylthiophene) (P3HT). The structure of these materials is shown in the inset of Figure 7.62a. It was intended that these would be present in the layer in equal proportions.

The purpose of the analysis was to confirm:

- That the two components of the organic layer were present in the expected ratio (1 : 1).
- That the ratio was constant throughout the layer
- That the elemental and chemical state composition was as expected
- That the valence band structure was indicative of the two components and was constant throughout the layer.

The XPS profile was used to determine the chemical composition of the layer while the UPS profile was used to examine the electronic structure of the VB.

Figure 7.62a shows the XPS profile through the layer and Figure 7.62b shows the same profile but with an expanded concentration scale. This profile shows that there are contaminants at the surface which are removed from the surface once ~50 nm of material has been removed. At depths greater than 50 nm the

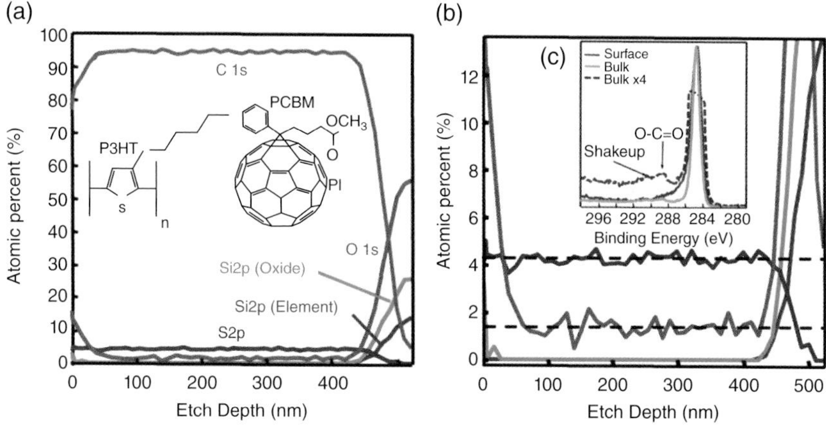


Figure 7.62 (a) The XPS depth profile through a layer of mixed phenyl-C61-butyric acid methyl ester (PCBM) and poly(3-hexylthiophene) (P3HT whose structures are also shown; (b) The same profile as that in (a) but with an expanded concentration scale. The dashed lines indicate the concentration of oxygen (4.2%) and sulphur (1.5%). (c) The C 1s spectrum from the surface of the layer (red) and from the bulk material (in green and expanded in blue).

composition of the film was shown to be constant until the interface with $SiO_2$ is reached. If $Ar^+$ ions had been used to sputter the material, there would have been a significant loss of O and S from the profile. As it is, the concentrations of these two elements remain constant at 4.2% (oxygen) and 1.5% (sulphur), these concentrations are indicated by the dashed lines in Figure 7.62b.

The sulphur in the profile should only come from the P3HT and the oxygen only from the PCBM. If that is the case, from the concentration of these two components in the steady state part of the profile, the concentrations of P3HT and PCBM can be calculated. In this case, these were found to be 46.6% P3HT and 53.2% PCBM, close to a ratio of 1 : 1. Note also that these two materials make up 99.8% of the layer suggesting both that the layer purity is very high and that there was little, if any, damage caused by sputtering.

The carbon chemistry near the surface of the organic layer is compared with that in the bulk of the layer in Figure 7.62c. The broad, almost featureless, peak (red spectrum) with the long tail on its high binding energy side is consistent with the presence of multiple types of contaminant. Once the contamination has been removed, clear features appear in the C 1s spectrum, the peak due to the O—C=O group and the shakeup peak from the aromatic rings both from the PCBM molecule. The presence of these peaks in the spectrum of the sputtered material is further evidence that sputtering did not damage the sample.

The UPS spectra, as a function of depth, are shown in Figure 7.63a and b. The montage, Figure 7.63a, clearly shows the constancy of the VB structure through-

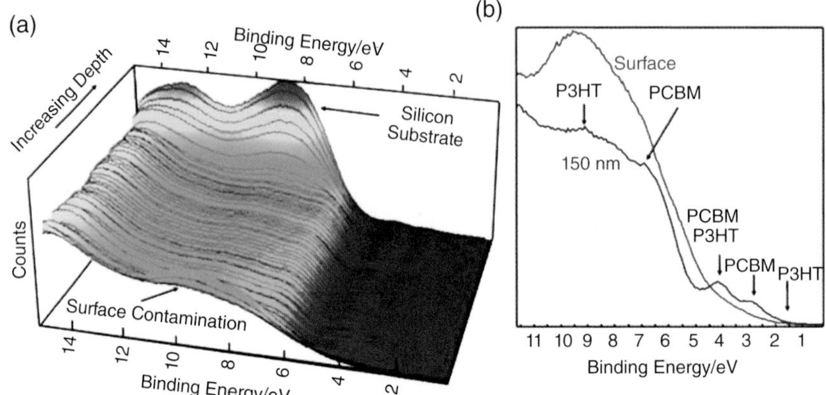


Figure 7.63 The ultra-violet photoelectron spectroscopy (UPS) spectra from the organic layer (a) a montage of all of the spectra acquired during the profile; (b) a spectrum from the surface compared with a representative spectrum from the bulk on which the features have been labelled.

out the organic layer, once the surface impurity layer has been removed. As would be expected from the XPS data, the VB only shows features associated with PCBM and P3HT when the contamination layer has been removed. The VB features labelled in Figure 7.63b are very similar to those found in the pure materials, further evidence that sputtering did not cause the layer to become damaged, and that there is no chemical interaction between the two components of the layer.

All of the analysis objectives have been met. This was achieved by profiling with argon cluster ions to maintain the integrity of the layer and by the use of both the XPS and the UPS data. The ability to obtain both types of data from the same profile is a major advantage because of both speed of analysis and confidence that the data are acquired from the same region of the sample.

While the examples cited in this chapter have dealt exclusively with the techniques described earlier in this book, it would be a very narrow-minded scientist who did not make full use of the plethora of advanced analytical techniques that are becoming available. In the next chapter, the more common ones are described and to enable comparisons to be drawn with XPS and AES.

# 8

# Comparison of XPS and AES with Other Analytical Techniques

XPS and AES are just two techniques which rely on the detection of electrons for the analysis of surfaces. There are several more. To those can be added techniques that rely upon the detection of photons and ions. We should also consider those techniques for which the input radiation is the photon, electron, and ion. Table 8.1 shows a list of the most popular of those techniques. The acronyms in this table are defined in the Glossary that follows.

Each of the techniques mentioned in this table has its own set of strengths and weaknesses and it is often necessary to use more than one technique to achieve a full analysis. For this reason, many surface analysis instruments can be used for two or more types of analysis.

The properties of some of the more popular techniques are summarised in Table 8.2.

*An Introduction to Surface Analysis by XPS and AES*, Second Edition.
John F. Watts and John Wolstenholme.

Table 8.1 Surface analysis techniques categorised by both the type of input radiation used and by the type of output radiation.

| In \ Out | Photons | Electrons | Ions |
|---|---|---|---|
| **Photons** | RAIRS (IRRAS)<br>Raman<br>Ellipsometry<br>Optical microscopy | XPS (ESCA)<br>UPS | MALDI |
| **Electrons** | EDS (EDX,)<br>WDS<br>IPES | AES<br>REELS<br>HREELS<br>EBSD<br>LEED<br>RHEED<br>SEM<br>TEM | ESD |
| **Ions** | PIXE | Ion-induced SEM<br>FIB imaging | SIMS<br>LEIS<br>MEIS<br>RBS |

*Source:* Reproduced by permission of Momentum Press, LLC, from 'Auger Electron Spectroscopy: Practical Application to Materials Analysis and Characterisation of Surfaces, Interfaces, and Thin Films', by John Wolstenholme, July 2015, all rights reserved.

Table 8.2 Features of various analytical methods.

| Technique | Acronym | Input radiation | Detected radiation | Analysis depth[a] | Lateral resolution[b] | Major advantages | Major disadvantages | Sensitivity[c] | Quantification | Chemical state information | Molecular structure | Vacuum required |
|---|---|---|---|---|---|---|---|---|---|---|---|---|
| Auger electron spectroscopy | AES | Electrons | Electrons | <10 nm | <10 nm | Spatial resolution | H not detected | ~0.1% | Yes but can be difficult | Limited | No | Yes |
| Ellipsometry | | Visible light | Visible light | | 1 μm | Thin film thickness measurement | No chemical or elemental information | N/A | Thickness only | No | No | No |
| High-resolution electron energy loss spectroscopy | HREELS | Electrons | Electrons | Adsorbed monolayer | 150 μm[d] | Structural information via the vibrational spectrum | Experimentally difficult | 0.1–1% of a monolayer | No | Yes | Via vibrational spectra | Yes |
| Inverse photoelectron spectroscopy | IPES | Electrons | Photons | 1 nm | 1 mm | Probes energy levels above the Fermi energy | Poor lateral resolution | | No | Electronic states | No | Yes |
| Low-energy electron diffraction | LEED | Electrons | Electrons | Top layer of atoms and adsorbed layer (kinematic LEED)<br>A few atomic layers probed in dynamic LEED | 100–500 μm | Crystal structure of the near-surface region. | Restricted to single-crystal samples | ~1% when LEED optics are used as an Auger spectrometer | Only when LEED optics are used as an Auger spectrometer | No | No | Yes |

(*Continued*)

Table 8.2 (Continued)

| Technique | Acronym | Input radiation | Detected radiation | Analysis depth[a] | Lateral resolution[b] | Major advantages | Major disadvantages | Sensitivity[c] | Quantification | Chemical state information | Molecular structure | Vacuum required |
|---|---|---|---|---|---|---|---|---|---|---|---|---|
| Low-energy ion scattering spectroscopy | LEISS | Ions | Ions | Top monolayer | >100 μm | Surface specificity | Poor mass resolution<br>Only sensitive to atoms heavier than incident ions | 1% (depends on sample) | No | No | No | Yes |
| Medium-energy ion scattering spectroscopy | MEISS | Ions | Ions | 40 nm | mm | Crystallographic information | Poor lateral resolution | 0.1% | No | No | No | Yes |
| Particle-induced X-ray emission | PIXE | Ions | X-rays | 5 μm | 100 μm | High sensitivity | Poor surface specificity | 10 ppm | Yes (with standards) | No | No | Yes |
| Reflection absorption infrared spectroscopy | RAIRS | Infrared | Infrared | Chemisorbed layer | ~15 μm | Structural information via the vibrational spectrum | No substrate information, no depth profiles | Variable depends on sample (few % of a monolayer) | No | Yes | Via vibrational spectra | No |
| Reflection electron energy loss spectroscopy[e] | REELS | Electrons | Electrons | <1 μm | 100 nm | Characterisation of valence bands and plasmons | No ability to identify elemental composition | | No | No | No | Yes |
| Rutherford back scattering | RBS | High-energy ions | High-energy ions | 1 μm (depth profile obtained) | >1 mm | Non-destructive depth profile | Large analysis area | $10^{14}$ atoms/$cm^2$<br>Dependent on mass of target atom | Yes | No | No | Not always |

| | | | | | | | | | | | | |
|---|---|---|---|---|---|---|---|---|---|---|---|---|
| Secondary ion mass spectrometry (dynamic) | d-SIMS | Ions | Ions | Depth profile up to ~10 μm | Depends on instrument type, typically <5 μm | High sensitivity and dynamic range<br>H detection | Quantification difficult in most cases | ppb | In some cases, with standards | No | No | Yes |
| Secondary ion mass spectrometry (static) | s-SIMS | Ions | Ions | <1 nm | <100 nm | High sensitivity<br>Molecular structure<br>H detection | Quantification not possible for most samples | ppm | No | Inferred from molecular structure | Yes | Yes |
| Ultraviolet photoelectron spectroscopy | UPS | UV light | Electrons | ~1 nm | >1 mm | Valence band structure determination | Not quantifiable | 1% | No | Electronic states | No | Yes |
| X-ray photoelectron spectroscopy | XPS | X-rays | Electrons | <10 nm | ~1 μm for imaging<br><10 μm for spectroscopy | Chemical state information | Lateral resolution limited to a few μm<br>H not detected | 0.1% | Yes | Yes | No | Yes |
| X-ray spectroscopy | EDS | Electrons | X-rays | ~1 μm | 0.5 μm | Quantification | Lack of surface specificity | 500 ppm | Yes | No | No | Yes |

*Source:* Reproduced by permission of Momentum Press, LLC, from 'Auger Electron Spectroscopy: Practical Application to Materials Analysis and Characterisation of Surfaces, Interfaces, and Thin Films', by John Wolstenholme, July 2015, all rights reserved.

[a] Without the use of sputtering, milling, and so on.

[b] Depending on the technique, this could be the spot size of the input radiation, the acceptance area of the analyser, or the resolution of an analytical image.

[c] In most cases, this is the detection limit, which can be achieved using a high-performance version of the instrumentation and when analysing a favourable sample.

[d] Likely to be different in $x$ and $y$ directions, 150 μm × 300 μm is a reasonable estimate.

[e] As performed using an Auger electron spectrometer.

# Glossary

Many of the terms and definitions are taken from ISO 18115 and are reproduced with the permission of the International Organization for Standardization, ISO. This standard can be obtained from any ISO member and from the website of the ISO Central Secretariat at the following address: http://www.iso.ch/ – now available FOC from NPL, NIST, etc.

## Abbreviations

| | |
|---|---|
| **AES** | Auger electron spectroscopy |
| **ALD** | atomic layer deposition |
| **ARXPS** | angle-resolved XPS (or angle-dependent XPS) |
| **CAE** | constant analyser energy |
| **CHA** | concentric hemispherical analysers |
| **CMA** | cylindrical mirror analyser |
| **CRR** | constant retard ratio |
| **DLD** | delay-line detector |
| **EBSD** | electron backscatter diffraction |
| **EELS** | electron energy loss spectroscopy |
| **ESCA** | electron spectroscopy for chemical analysis |
| **EDX (EDS)** | energy dispersive X-ray spectroscopy (or spectrometer) |
| **EPMA** | electron probe micro analysis |
| **FAT** | fixed analyser transmission |
| **FIB** | focused ion beam |
| **FRR** | fixed retard ratio |
| **FWHM** | full width at half maximum |
| **GO** | graphene oxide |
| **HAXPES** (or **HAXPS**) | hard X-ray photoelectron spectroscopy |

*An Introduction to Surface Analysis by XPS and AES*, Second Edition.
John F. Watts and John Wolstenholme.

| | |
|---|---|
| **HPLC** | High-performance liquid chromatography. |
| **HSA** | hemispherical sector analyser |
| **IMFP** | inelastic mean free path |
| **ISS** | ion scattering spectroscopy |
| **LEISS** | low-energy ion scattering spectroscopy |
| **LMIG** | liquid metal ion gun |
| **MALDI** | matrix-assisted laser desorption/ionisation mass spectrometry |
| **MCP** | microchannel plate |
| **NAPXPS** | near ambient pressure XPS |
| **NPA** | normalised peak area |
| **OFET** | organic field-effect transistor |
| **OLED** | organic light-emitting diode |
| **PA** | peak area |
| **PET** | polyethylene terephthalate |
| **PSD** | position-sensitive detector |
| **ptp** | peak-to-peak |
| **QUASES** | quantitative analysis of surfaces by electron spectroscopy |
| **RAD** | resistive-anode detector |
| **RBS** | Rutherford backscattering spectrometry |
| **RDP** | relative depth plot |
| **REELS** | reflection electron energy-loss spectroscopy |
| **RHEED** | reflection high-energy electron diffraction |
| **SAM** | scanning Auger microscope (or microscopy) |
| **SAM** | self-assembled monolayer |
| **SAXPS (SAX)** | small area XPS |
| **SEM** | scanning electron microscope |
| **SIMS** | secondary ion mass spectrometry |
| **SMA** | spherical mirror analyser |
| **SNMS** | sputtered neutral mass spectrometry |
| **SSA** | spherical sector analyser |
| **TOF (ToF)** | time of flight |
| **UHV** | ultra-high vacuum |
| **UPS** | ultra-violet photoelectron spectroscopy |
| **VPSEM** | variable pressure SEM |
| **WDX (WDS)** | wavelength-dispersive X-ray spectroscopy (or spectrometer) |
| **X-AES** | X-ray induced Auger electron spectroscopy |
| **XPS** | X-ray photoelectron spectroscopy |
| **XRD** | X-ray diffraction |

## Surface Analysis Methods

**Auger electron spectroscopy** a method in which an electron spectrometer is used to measure the energy distribution of Auger electrons emitted from a surface.

**Electron Spectroscopy for Chemical Analysis (ESCA)** a method encompassing both AES and XPS. The term ESCA is falling out of use as, in practice, it was only used to describe situations more clearly defined by the term X-ray photoelectron spectroscopy (XPS). The latter term has been preferred.

**Secondary Ion Mass Spectrometry (SIMS)** a method in which a mass spectrometer is used to measure the mass-to-charge quotient and abundance of secondary ions emitted from a sample as a result of bombardment by energetic ions.

**Ultra-violet Photoelectron Spectroscopy (UPS)** a method in which an electron spectrometer is used to measure the energy distribution of photoelectrons emitted from a surface irradiated by ultra-violet photons. Ultra-violet sources in common use include various types of discharges that can generate the resonance lines of various gases (e.g. the He I and He II emission lines at energies of 21,2 and 40,8 eV, respectively). For variable energies, synchrotron radiation is used.

**X-ray Photoelectron Spectroscopy (XPS)** a method in which an electron spectrometer is used to measure the energy distribution of photoelectrons and Auger electrons emitted from a surface irradiated by X-ray photons. X-ray sources in common use are Al and Mg non-monochromated K$\alpha$ X-rays at 1486,6 and 1253,6 eV, respectively. Modern instruments also use monochromated Al K$\alpha$ X-rays. Some instruments make use of various X-ray sources with other anodes or of synchrotron radiation.

## Terms Used in Surface Analysis

**Adventitious Carbon Referencing** determining the charging potential of a particular sample from a comparison of the experimentally determined C 1s binding energy, arising from adsorbed hydrocarbons on the sample, with a standard binding energy value. A nominal value of 285.0 eV is often used for the binding energy of the relevant C 1s peak, although some analysts prefer specific values in the range 284.6–285.2 eV depending on the nature of the substrate.

**Angle Lapping** sample preparation in which a sample is mechanically polished at an angle to the original surface. This angle may often be less than 1° so that depth information with respect to the original surface is transformed to lateral information.

**Angle of Emission** angle between the trajectory of a particle or photon as it leaves a surface and the local or average surface normal.

**Angle of Incidence** angle between the incident beam and the local or average surface normal.

**Angle-resolved XPS (ARXPS) or angle-dependent XPS** a procedure in which X-ray photoelectron intensities are measured as a function of the angle of emission.

**Angle, Take-off** angle between the trajectory of a particle as it leaves a surface and the normal to the local or average surface plane.

**Attenuation Length** quantity $\lambda$ in the expression $\Delta x/\lambda$ for the fraction of a parallel beam of specified particles or radiation removed in passing through a thin layer $\Delta x$ of a substance in the limit as $\Delta x$ approaches zero, where $\Delta x$ is measured in the direction of the beam. The intensity or number of particles in the beam decays as $\exp(-x/\lambda)$ with the distance $x$.

**Auger Electron** electron emitted from atoms in the Auger process.

**Auger Electron Spectrum** plot of the Auger electron intensity as a function of the electron kinetic energy, usually as part of the energy distribution of detected electrons.

**Auger Parameter** kinetic energy of a narrow Auger electron peak in a spectrum minus the kinetic energy of the most intense photoelectron peak from the same element.

**Auger Process** relaxation, by electron emission, of an atom with a vacancy in an inner electron shell, the emitted electrons have characteristic energies, defined by the Auger transition.

**Auger Transition** Auger process involving designated electron shells or sub-shells. The three shells involved in the Auger process are designated by three letters. The first letter designates the shell containing the initial vacancy and the last two letters designate the shells containing electron vacancies left by the Auger process (for example, KLL, and LMM). When a valence electron is involved the letter V is used (for example, LMV and KVV). When a particular sub-shell involved is known this can also be indicated (for example, $KL_1L_2$). Coupling terms may also be added, where known, to indicate the final atomic state (for example, $L_3M_{4,5}M_{4,5};{}^1D$).

**Background, Inelastic** intensity distribution in the spectrum for particles originally at one energy but which are emitted at lower energies due to one or more inelastic scattering processes.

**Backscattered Electron** electron originating in the incident beam which is emitted after interaction with the sample.

**Ball Cratering** a procedure in which the sample is abraded by a sphere in order to expose compositional changes in layers below the original surface with the intent that the depth of those layers can be related to the lateral position in the crater created by the abrasion.

**Beam Current** quotient of d$Q$ by d$t$, where d$Q$ is the quantity of electric charge of a specified polarity in the beam passing in the time interval d$t$

$I = \mathrm{d}Q/\mathrm{d}t$

**Beam Current Density** quotient of d$I$ by d$A$ where d$I$ is the element of beam current incident on an area d$A$ at right angles to the direction of the beam

$J = \mathrm{d}I/\mathrm{d}A$

**Beam Diameter** for a particle beam of circular cross section full width of the beam at half maximum intensity measured in a plane normal to the beam direction.

**Binding Energy** energy that must be expended in removing an electron from a given electronic level to the Fermi level of a solid or to the vacuum level of a free atom or molecule.

**Bremsstrahlung** photon radiation emitted from a material due to the deceleration of incident electrons within that material. The bremsstrahlung radiation has a continuous spectral distribution up to the energy of the incident electrons. In XPS, The bremsstrahlung from a conventional X-ray source with an Al or Mg anode leads to a continuous photoelectron background. This radiation may also photoionise inner shells that would be energetically impossible by characteristic Al or Mg K$\alpha$ X-rays. As a result, Auger electron features may appear at negative binding energy values and, in addition, the intensities of other Auger electron features may be greater than if the inner shell vacancies had been created only by the characteristic X-rays. The bremsstrahlung-excited Auger electron features can be helpful for determining the various Auger parameters needed to identify chemical states.

**Chemical Shift** change in peak energy arising from a change in the chemical environment of the atom.

**Compositional Depth Profile** chemical or atomic composition measured as a function of distance normal to the surface.

**Constant ΔE Mode (constant analyser energy mode, CAE mode or fixed analyser transmission mode, FAT mode)** mode of electron energy analyser operation that varies the electron retardation but keeps the pass energy constant in the energy dispersive portion of the analyser. This mode is often used in XPS to maintain a high and constant energy resolution throughout the spectrum.

**Constant ΔE/E Mode (constant retardation ratio mode, CRR mode or fixed retardation ratio mode, FRR mode)** mode of electron energy analyser operation that varies the retarding potential so that the pass energy in the energy dispersive portion of the analyser is a constant fraction of the original **vacuum level** referenced kinetic energy. This mode is often used in AES to improve the signal-to-noise ratio for high energy emitted electrons at the expense of spectral resolution.

**Depth Profiling** monitoring of signal intensity as a function of a variable that can be related to distance normal to the surface usually, sputtering time.

**Depth Resolution** depth range over which a signal changes by a specified quantity when reconstructing the profile of an ideally sharp interface between two media or a delta layer in one medium. The precise quantity to be used depends on the signal function with depth. However, for routine analytical use, a convention of the depth at an interface over which the signal from an overlayer or a substrate changes from 16 to 84% of their total variation between plateau values, is often used in AES and XPS.

**Detection Limit** smallest amount of an element or compound that can be measured under specified analysis conditions.

**Differential Spectrum** differential of the direct spectrum with respect to energy, *E*, by an analogue electrode modulation method or by numerical differentiation of that spectrum.

**Direct Spectrum** intensity of electrons transmitted and detected by a spectrometer with a dispersing energy analyser, as a function of energy *E*.

**Electron Retardation** method of measuring the kinetic energy distribution by retarding the emitted electrons before or within the electron energy analyser.

**Grazing Exit (or Glancing Exit)** geometrical arrangement in which the angle of the scattered (or emitted) particles is near 90° from the normal to the sample surface. This configuration generally results in improved surface sensitivity and may also improve depth resolution.

**Grazing Incidence (or Glancing Incidence)** geometrical arrangement in which the angle of the incident particles is near 90°° from the normal to the sample surface.

**Inelastic Mean Free Path, Electron** average distance that an electron with a given energy travels between successive inelastic collisions.

**Inelastic Scattering** interaction between a moving energetic particle and a second particle or assembly of particles in which the total kinetic energy is not conserved. Kinetic energy is absorbed in solids by various mechanisms, for example, inner shell ionisation, plasmon and phonon excitation and bremsstrahlung generation. These excitations usually lead to a small change in direction of the moving particle.

**Information Depth** maximum depth, normal to the surface, from which useful information is obtained.

**Interface Width, Observed** distance over which a 16–84%, or 84–16%, change in signal intensity is measured at the junction of two dissimilar matrices, the thicknesses of which are more than six times that distance.

**Lateral Resolution** distance measured either in the plane of the sample surface or in a plane at right angles to the axis of the image-forming optics over which changes in composition can be separately established with confidence.

**Line Scan** plot of the output signal intensity from the spectrometer, the signal intensity from another detector, or processed intensity information from the available software along a line corresponding to a line on the sample surface.

**Map or Image** two-dimensional representation of the sample surface where the information at each point in the representation is related to the output signal from the spectrometer, the signal from another detector, or processed intensity information from the available software. By convention, map is usually applied to cases where the information is primarily composition-specific and image to those where it is primarily topographic. Map intensities may be presented in a normalised fashion to have the maximum and minimum signal intensities set at, for example, full white and full black, respectively, or on a colour scale. The contrast scale should be defined.

**Monolayer** complete coverage of a substrate by one atomic or molecular layer of a species.

**Multiplet Splitting** splitting of an Auger electron line into two or more components caused by the interactions of the atomic vacancies created by the Auger process.

or

splitting of a photoelectron line caused by the interaction of the unpaired electron created by photoemission with other unpaired electrons in the atom.

**Noise** time varying disturbances superimposed on the analytical signal with fluctuations leading to uncertainty in the signal intensity.

**Noise, Statistical** noise in the spectrum due solely to the statistics of randomly detected single events. For single-particle counting systems exhibiting Poisson statistics, the standard deviation of a large number of measures of an otherwise steady count rate, $N$, each in the same time interval, is equal to the square root of $N$.

**Pass Energy** means kinetic energy of the detected particles in the energy dispersive portion of the energy analyser.

**Peak Fitting** a procedure whereby a spectrum, generated by peak synthesis, is adjusted to match a measured spectrum. A least squares optimisation procedure is generally used in a computer program for this purpose.

**Peak Synthesis** procedure whereby a synthetic spectrum is generated using either model or experimental peak shapes in which the number of peaks, the peak shapes, peak widths, peak positions, peak intensities and the background shape and intensity are adjusted for peak fitting.

**Peak-to-Background Ratio (or Signal-to-Background Ratio)** ratio of the maximum height of the peak above the background intensity to the magnitude of that background intensity.

**Peak Width** width of a peak at a defined fraction of the peak height. The most common measure of peak width is the full width of the peak at half maximum (FWHM) intensity.

**Photoelectric Effect** interaction of a photon with bound electrons in atoms, molecules, and solids, resulting in the production of one or more photoelectrons.

**Photoelectron X-ray Satellite Peaks** photoelectron peaks in a spectrum resulting from photoemission induced by characteristic minor X-ray lines associated with the X-ray spectrum of the anode material.

**Photoemission** emission of electrons from atoms or molecules caused by the photoelectric effect.

**Plasmon** excitation of valence band electrons in a solid in which collective oscillations are generated.

**Primary Electron** electron extracted from a source and directed to a sample.

**Profile, Depth** chemical or elemental composition, signal intensity, or processed intensity information from the available software measured in a direction normal to the surface.

**Raster** two-dimensional pattern generated by the deflection of a primary beam. Commonly used rasters cover square or rectangular areas.

**Relative Resolution of a Spectrometer** ratio of the resolution of a spectrometer at a given energy, mass, or wavelength to that energy, mass, or wavelength.

**Secondary Electron** electron, generally of low energy, leaving a surface as a result of an excitation induced by an incident electron, photon, ion, or neutral particle.

**Selected Area Aperture** aperture in the electron or ion optical system restricting the detected signal to a small area of the sample surface.

**Shakeup** multielectron process in which an atom is left in an excited state following a photoioization or Auger electron process, so that the outgoing electron has a characteristic kinetic energy slightly less than that of the parent photoelectron.

**Signal-to-Noise Ratio** ratio of the signal intensity to a measure of the total noise in determining that signal.

**Smoothing** mathematical treatment of data to reduce the apparent noise.

**Spectrometer Transmission Function** quotient of the number of particles transmitted by the analyser by the number of such particles per solid angle and per interval of the dispersing parameter (e.g. energy, mass, or wavelength) available for measurement as a function of the dispersing parameter.

**Spin Orbit Splitting** splitting of p, d, or f levels in an atom arising from coupling of the spin and orbital angular momentum.

**Sputter Depth Profile** compositional depth profile obtained when the surface composition is measured as material is removed by sputtering.

**Sputtering** process in which atoms and ions are ejected from the sample as a result of particle bombardment.

**Sputtering, Preferential** change in the equilibrium surface composition of the sample which may occur when sputtering multicomponent samples.

**Sputtering Rate** quotient of the amount of sample material removed, as a result of particle bombardment, by time.

**Sputtering Yield** ratio of the number of atoms and ions sputtered from a sample to the total number of incident primary particles.

**Vacuum Level** electric potential of the vacuum at a point in space. In electron spectroscopy, the point in space is taken at a sufficiently large distance outside the sample such that electric fields caused by different work functions of different parts of the surface are zero or extremely small.

**Vacuum Level Referencing** method of establishing the kinetic energy scale in which the zero point corresponds to an electron at rest at the vacuum level.

**Valence Band Spectrum** photoelectron energy distribution arising from excitation of electrons from the valence band of the sample material.

**Work Function** potential difference for electrons between the Fermi level and the maximum potential just outside a specified surface.

**X-ray monochromator** device used to eliminate photons of energies other than those in a narrow energy or wavelength band.

# Bibliography

## Chapter 1

Briggs, D. and Seah, M.P. (1990). *Practical Surface Analysis by Auger and X-ray Photoelectron Spectroscopy*, 2e. Chichester, UK: Wiley.

Briggs, D. and Grant, J.T. (2003). *Surface Analysis by Auger and X-ray Photoelectron Spectroscopy*. Chichester, West Sussex, UK: IM Publications and SurfaceSpectra Ltd.

Hofmann, S. (2013). *Auger and X-ray Photoelectron Spectroscopy in Materials Science: A User-oriented Guide*. Berlin Heidelberg, Germany: Springer-Verlag.

NIST X-ray Photoelectron Spectroscopy Database (2000). NIST Standard Reference Database Number 20, National Institute of Standards and Technology, Gaithersburg MD. doi:https://doi.org/10.18434/T4T88K

Prutton, M. and El Gomati, M.M. (2006). *Scanning Auger Electron Microscopy*. Chichester, UK: Wiley.

Riviere, J.C. and Myra, S. (2009). *Handbook of Surface and Interface Analysis: Methods for Problem Solving*. Roca Baton, FL, USA: CRC Press Taylor & Francis Group.

Seah, M.P. and Dench, W.A. (1979). Quantitative electron spectroscopy of surfaces: a standard data base for electron inelastic mean free paths in solids. *Surf. Interface Anal.* 1: 2–11.

Seah, M.P. (2012). An accurate and simple universal curve for the energy-dependent electron inelastic mean free path. *Surf. Interface Anal.* 44: 497–503.

Seah, M.P. (2012). Simple universal curve for the energy-dependent electron attenuation length for all materials. *Surf. Interface Anal.* 44: 1353–1359.

Seighbahn, K., Fahlman, A., and Nordling, C. (1967). *ESCA: Atomic, Molecular, and Solid State Structure Studied by Means of Electron Spectroscopy*. Uppsala, Sweeden: Almqvist and Wiksells.

van der Heide, P. (2012). *X-ray Photoelectron Spectroscopy: An Introduction to Principles and Practices*. Hoboken, New Jersey, USA: Wiley.

*An Introduction to Surface Analysis by XPS and AES*, Second Edition.
John F. Watts and John Wolstenholme.

Vickerman, J.C. and Gilmore, I.S. (2009). *Surface Analysis: The Principal Techniques*. Chichester, UK: Wiley.

Wolstenholme, J. (2015). *Auger Electron Spectroscopy: Practical Applications to Materials Analysis and Characterization of Surfaces, Interfaces and Thin Films*. New York, USA: Momentum Press LLC.

## Chapter 2

Brooker, A.D. and Castle, J.E. (1986). Scanning Auger microscopy: resolution in time, energy and space. *Surf. Interface Anal.* 8: 113–119.

Castle, J.E. and West, R.H. (1980). The utility of bremsstralung induced Auger peaks. *J. Elec. Spec.* 18: 355–358.

Dietrich, P.M., Bahr, S., Yakamoto, T. et al. (2019). Chemical analysis on materials and devices under functional conditions – Environmental photoelectron spectroscopy as a non-destructive tool for routine characterization. *J. Elec. Spec.* 231: 118–126.

Edgell, M.J., Paynter, R.W., and Castle, J.E. (1985). High energy XPS using a monochromated AgL$\alpha$ source: resolution, sensitivity and photoelectric cross sections. *J. Elec. Spec.* 37: 241–256.

Fairley N, Carrick A, (2005). The CASA Cookbook: Recipes for XPS Data Processing, Pt. 1. Teignmouth, Devon UK. www.casaxps.com.

Fairley N, Walton J, (2011). The CASA Cookbook: XPS Imaging Processing, Pt 2: The Casa XPS User's Manual. Teignmouth, Devon, UK. www.casaxps.com

Kobayashi, K., Kobata, M., and Iwai, H. (2013). Development of a laboratory system hard XPS and its applications. *J. Elec. Spec.* 190: 210–221.

Regoutz, A., Mascheck, M., Wiell, T. et al. (2018). A novel laboratory-based hard X-ray photoelectron spectroscopy system. *Rev. Sci. Instrum.* 89: 073105.

Wagner, C.D. and Joshi, A. (1988). The Auger parameter, its utility and advantages: a review. *J. Elec. Spec.* 47: 283–313.

Weiland, C., Rumaiz, A.K., Pianetta, P. et al. (2016). Recent applications of hard XPS. *J. Vac. Sci. Technol. A* 34 (3): 1–21.

## Chapter 3

Childs, K.D., Carlson, B.A., LaVanier, L.A. et al. (1995). *Handbook of Auger Electron Spectroscopy*. Eden Prairie, USA: Physical Electronics Division, Physical Electronics Inc.

Hall, P.M. and Morabito, J.M. (1979). Matrix effects in the quantitative Auger analysis of dilute alloys. *Surf. Sci.* 83: 391–405.

Hall, P.M., Morabito, J.M., and Conley, D.K. (1977). Relative sensitivity factors for Auger analysis of binary alloys. *Surf. Sci.* 62: 1–20.

Kaciulis, S. (2012). Spectroscopy of carbon: from diamond to nitride films. *Surf. Interface Anal.* 44: 1155–1161.

Kaciulis, S., Mezzi, A., Calvani, P., and Trucchi, D.M. (2014). Electron spectroscopy of the main allotropes of carbon. *Surf. Interface Anal.* 46: 966–969.

Moulder, J.F., Stickle, W.F., Sobol, P.E. et al. (1995). *Handbook of X-ray Photoelectron Spectroscopy*. Eden Prairie, USA: Physical Electronics Division, Perkin-Elmer Corporation.

Seah, M.P. (1979). Quantitative AES: via the energy spectrum of the differential? *Surf. Interface Anal.* 1: 86–90.

Seah, M.P. (1980). The quantitative analysis of surfaces by XPS: a review. *Surf. Interface Anal.* 2: 222–239.

Seah M P, (1983). Review of quantitative Auger electron spectroscopy. Scanning Electron Microscopy, SEM Inc., Chicago, USA, 521–36.

Sherwood, P.M.A. (2019). The use and misuse of curve fitting in the analysis of core X-ray photoelectron spectroscopic data. *Surf. Interface Anal.* 1–22. https://doi.org/10.1002/sia.6629.

Turgeon, S. and Paynter, R.W. (2001). On the determination of carbon $sp^2/sp^3$ ratios in polystyrene/polyethylene copolymers by photoelectron spectroscopy. *Thin Solid Films* 394: 44–48.

Wagner, C.D., Davis, L.E., Zeller, M.V. et al. (1981). Empirical sensitivity factors for quantitative analysis by electron spectroscopy for chemical analysis. *Surf. Interface Anal.* 3: 211–225.

## Chapter 4

Chang, J.P., Green, M.L., Donnelly, V.M. et al. (2000). Profiling nitrogen in ultrathin silicon oxynitrides with angle resolved X-ray photoelectron spectroscopy. *J. Appl. Phys.* 87: 4449.

Cumpson, P.J. (1995). Angle resolved XPS and AES: depth resolution limits and a general comparison of depth profile reconstruction methods. *J. Electron Spectrosc. Relat. Phenom.* 73: 25.

Kelly, R. (1985). On the role of gibbsian segregation in causing preferential sputtering. *Surf. Interface Anal.* 7: 1–7.

Lea, C. and Seah, M.P. (1981). Optimized depth resolution in ion-sputtered and lapped compositional profiles with Auger electron spectroscopy. *Thin Solid Films* 75: 67–86.

Seah, M.P. (1981). Pure element sputtering yields using 500-1000 eV argon ions. *Thin Solid Films* 81: 279–287.

Seah, M.P. and Hunt, C.P. (1983). The depth dependence of the depth resolution in composition-depth profiling with Auger electron spectroscopy. *Surf. Interface Anal.* 5: 33–37.

Smith, G.C. and Livsey, A.K. (1992). Maximum entropy: a new approach to non-destructive depth from angle dependent XPS. *Surf. Interface Anal.* 19: 175.
Tougaard, S. (2018). Improved XPS analysis by visual inspection of the survey spectrum. *Surf. Interface Anal.* 50: 657–666.
Walls, J.M., Hall, D.D., and Sykes, D.E. (1979). Compositional-depth profiling and interface analysis of surface coatings using ball-cratering and the scanning Auger microprobe. *Surf. Interface Anal.* 1: 204–210.
Yih, R.S. and Ratner, B.D. (1987). A comparison of two angular dependent ESCA algorithms useful for constructing depth profiles for surfaces. *J. Elec. Spec.* 43: 61–82.

## Chapter 5

Flewitt, P.E.J. and Wild, R.K. (2017). *Physical Methods for Materials Characterisation.* Boca Raton, FL, USA: Taylor & Francis Group/CRC Press.
Sherwood, P.M.A. (2003). Valence bands studied by XPS. In: *Surface Analysis by Auger and X-ray Photoelectron Spectroscopy* (eds. D. Briggs and J.T. Grant), 531–555. Chichester, West Sussex, UK: IM Publications and SurfaceSpectra Ltd.

## Chapter 6

Baer, D.R., Lea, A.S., Cazaux, J. et al. (2010). Approaches to analysing insulators with Auger electron spectroscopy: Update and overview. *J. Elec. Spec.* 176: 80–94.
Edwards, L., Mack, P., and Morgan, D.J. (2019). Recent advances in dual mode charge compensation for XPS. *Surf Interface Anal* 51 (9): 925–933. DOI: 10.1002/sia.6680.

## Chapter 7

Aeimbhu, A., Castle, J.E., and Singhai, P. (2005). Accounting for the size of molecules in determination of adsorption isotherms by XPS; exemplified by adsorption of chicken egg albumin on titanium. *Surf. Interface Anal.* 37: 1127–1136.
Baer, D.R. and Engelhard, M.H. (2010). XPS analysis of nanostructured materials and biological surfaces. *J. Electron Spectrosc. Relat. Phenom.* 178-179: 415–432.
Baer, D.R., Gaspar, D.J., Nachimuthu, P. et al. (2010). Application of surface chemical analysis tools for characterization of nanoparticles. *Anal. Bioanal. Chem.* 396: 983–1002.
Baker, M.A. and Castle, J.E. (1992). The initiation of pitting corrosion of stainless steels at oxide inclusions. *Corros. Sci.* 33: 1295–1312.

Baker, M.A. and Castle, J.E. (1993). The initiation of pitting corrosion at MnS inclusions. *Corros. Sci.* 34: 667–682.

Baker, M.A. (1995). AES/XPS study of Ti-B-N thin films. *J. Vac. Sci. Technol.* A13: 1633–1638.

Baker, M.A., Mollart, T.P., Gibson, P.N. et al. (1997). Combined XPS/AES/GXRD/EXAFS Investigation of $TiB_xN_y$ Coatings. *J. Vac. Sci. Technol.* A15: 284–291.

Baker, M.A., Gilmore, R., Lenardi, P.N. et al. (1999). Microstructure and mechanical properties of multilayer $TiB_2$-C and co-sputtered $TiB_2$-C coatings for cutting tools. *Vacuum* 53: 113–116.

Beamson, G. and Briggs, D. (1992). *High Resolution XPS of Organic Polymers: The Scienta ESCA300 Database*. Chichester, UK: Wiley.

Bhasin, M.M. (1975). Auger spectroscopic analysis of the poisoning of commercial palladium-alumina hydrogenation catalyst. *J. Catal.* 38: 218–222.

Breeson, A., Sankar, G., Goh, G.K.L. et al. (2017). Phase quantification by X-ray photoemission valence band analysis applied to mixed phase $TiO_2$ powders. *Appl. Surf. Sci.* 423: 205–209.

Brinen, J.S., Graham, S.W., Hammond, J.S. et al. (1984). Characterization of fresh and spent HDS catalysts by Auger and X-ray photoelectron spectroscopies. *Surf. Interface Anal.* 6: 68–73.

Castle, J.E. (1986). The role of electron spectroscopy in corrosion science. *Surf. Interface Anal.* 9: 345–356.

Castle, J.E. (2018). The application of electron spectroscopy to studies of passivity of metals and alloys. In: *Encyclopedia of Interfacial Chemistry: Surface Science and Electrochemistry*, vol. vol. 6 (ed. K. Wandelt), 469–477. Elsevier Inc.

Cumpson, P.J. (2000). The Thickogram: a new method for easy film thickness measurement in XPS. *Surf. Interface Anal.* 29: 403–406.

Diplas, S., Watts, J.F., Tsakiropoulos, P. et al. (2001). XPS studies of Ti-Al and Ti-Al-V alloys using CrKβ radiation. *Surf. Interface Anal.* 31: 734–744.

Grilli, R., Baker, M.A., Castle, J.E. et al. (2011). Corrosion behaviour of a 2219 aluminium alloy treated with a chromate conversion coating exposed to a 3.5% NaCl solution. *Corros. Sci.* 53: 1214–1233.

Haupt, J., Baker, M.A., Stroosnijder, M.F. et al. (1994). AES studies on $TiN_x$. *Surf. Interface Anal.* 22: 167–170.

Jesson, D.A. and Watts, J.F. (2012). The interface and interphase in polymer matrix composites: effect on mechanical properties and methods for identification. *J. Macromol. Sci. Polym. Rev.* 52: 321–354.

Leadley, S.R. and Watts, J.F. (1997). *J. Adhes.* 60: 175–196.

Mallinson, C.F., Yates, P.M., Baker, M.A. et al. (2017). The localised corrosion associated with second phase particles in AA7075-T6: a study by SEM, EDX, AES, SKPFM and FIB-SEM. *Mater. Corros.* 68: 748–763.

Morgan, D.J. (2019). Imaging XPS for Industrial Applications. *J Elec Spec Rel Phenom* 231: 109–117.

O'Hare, L.-A., Smith, J.A., Leadley, S.R. et al. (2002). Surface physico-chemistry of Corona-discharge-treated PET film. *Surf. Interface Anal.* 33: 617–625.

Paynter, R.W. and Ratner, B.D. (1985). The study of interfacial proteins and bimolecules by X-ray photoelectron spectroscopy. In: *Surface and Interfacial Aspects of Biomedical Polymers* (ed. J.D. Andrade), 189–216. New York: Plenum Press.

Perruchot, C., Abel, M.-L., Watts, J.F. et al. (2002). High resolution XPS of crosslinking and segregation phenomena in hexamethoxymethyl melamine polyester resins. *Surf. Interface Anal.* 34: 570–574.

Perruchot, C., Watts, J.F., Lowe, C. et al. (2002). Angle-resolved XPS characterisation of urea formaldehyde-epoxy systems. *Surf. Interface Anal.* 33: 869–878.

Pijpers, A.P. and Meier, R.J. (1987). Oxygen-induced secondary substituent effects in polymer XPS spectra. *J. Elec. Spec.* 43: 131–137.

Prickett, A.C., Smith, P.A., and Watts, J.F. (2001). ToF-SIMS studies of carbon fibre fracture surfaces and the development of controlled Mode in situ fracture. *Surf. Interface Anal.* 31: 11–17.

Riviere, J.C. and Myra, S. (2009). *Handbook of Surface and Interface Analysis: Methods for Problem Solving.* Roca Baton, FL, USA: Taylot & Francis Group/ CRC Press.

Rouxhet, P.G. and Genet, M.J. (2011). XPS analysis of bio-organic systems. *Surf. Interface Anal.* 43: 1453–1470.

Seah, M.P. (1980). *J. Vac. Sci. Technol.* 17: 16.

Schwartz, V., Fu, W., Tsai, Y.-T. et al. (2013). Oxygen-functionalized few-layer graphene sheets as active catalysts for oxidative dehydrogenation reactions. *ChemSusChem* 6: 840–846.

Seah, M.P. and Hondros, E.D. (1977). Segregation to interfaces. *Int. Met. Rev.* 22: 262–301.

Shard, A.G. (2012). A straightforward method for interpreting XPS data from core-shell nanoparticles. *J. Phys. Chem. C* 116: 16806–16813.

Shard, A.G., Wang, J., and Spencer, S.J. (2009). XPS topofactors: determining overlayer thicknesses on particles and fibres. *Surf. Interface Anal.* 41: 541–548.

Shard, A.G., Havelund, R., Spencer, S.J. et al. (2015). Measuring compositions in organic depth profiling: results from a VAMAS interlaboratory study. *J. Phys. Chem. B* 119: 10784–10797.

Tardio, S., Abel, M.L., Carr, R.H. et al. (2015). Comparative study of the native oxide on 316L stainless steel by XPO and ToF-SIMS. *J. Vac. Sci. Technol.* A33: 1–14.

The Proceedings of the Biennial European Conference on Applications of Surface and Interface Analysis (ECASIA) is published as a single bound volume of Surface and Interface Analysis. These proceedings provide a timely overview of the application of surface analysis in all aspects of materials science.

Trindade, G.F., Abel, M.-L., and Watts, J.F. (2017). Non-negative matrix factorisation (NMF) of large mass spectrometry data sets. *Chemom. Intell. Lab. Syst.* 163: 76–85.

Watts, J.F. (1985). Analysis of ceramic materials by electron spectroscopy. *J. Microsc.* 140: 243–260.

Watts, J.F. (2009). Adhesion science and technology. In: *Handbook of Surface and Interface Analysis: Methods for Problem Solving* (eds. J.C. Riviere and S. Myhra), 5651–5602. Boca Raton, FL, USA: CRC Press an imprint of Taylor and Francis Group.

Watts, J.F. (2018). The use of surface analysis methods to probe the interfacial chemistry of adhesion. In: *Handbook of Adhesion Technology*, 2e, vol. 1 (eds. L.M.F. da Silva, A. Öchsner and R.D. Adams), 227–255. Heidelberg Germany: Springer.

Watts, J.F. and Castle, J.E. (1999). The determination of adsorption isotherms by XPS and ToF-SIMS: their role in adhesion science. *Int. J. Adhes. Adhes.* 19: 435–443.

Watts, J.F., Abel, M.-L., Perruchot, C. et al. (2001). Segregation and crosslinking in urea formaldehyde resins: a study by high resolution XPS. *J. Elec. Spec.* 121: 233–247.

West, R.H. and Castle, J.E. (1982). The correlation of the Auger parameter with refractive index: and XPS study of silicates using ZrLα radiation. *Surf. Interface Anal.* 4: 68–75.

Williams, D.F., Abel, M.-L., Grant, E. et al. (2015). Flame treatment of polypropylene: a study by electron and ion spectroscopies. *Int. J. Adhes. Adhes.* 63: 26–33.

Wu, J., Song, S.-H., Weng, L.-Q. et al. (2008). An Auger electron spectroscopy study of phosphorus and molybdenum grain boundary segregation in a Cr2.25Mo1 steel. *Mater. Charact.* 59: 261–265.

## Chapter 8

Brongersma, H.H., Draxler, M., de Ridder, M., and Bauer, P. (2007). Surface composition analysis by low-energy ion scattering. *Surf. Sci. Rep.* 62: 63–109.

Brydson, R.M.D. (2001). *Electron Energy Loss Spectroscopy*. Oxford, UK: BIOS Scientific Publishers.

Budd, P.M. and Goodhew, P.J. (1988). *Light-element Analysis in the Transmission Electron Microscope: WEDX and EELS*. Oxford: Oxford University Press.

Castle, J.E. and Castle, M.D. (1983). Simultaneous XRF and XPS analysis. *Surf. Interface Anal.* 5: 193–198.

Chu, W.K., Mayer, J.W., and Nicolet, M.-A. (1978). *Backscattering Spectrometry*. New York: Academic Press Inc.

Clarke, N.S., Ruckman, J.C., and Davey, A.R. (1986). The application of laser ionization mass spectrometry to the study of thin films and near-surface layers. *Surf. Interface Anal.* 9: 31–40.

Degreve, F., Thorne, N.A., and Lang, J.M. (1988). Metallurgical applications of SIMS. *J. Mater. Sci.* 23: 4181–4208.

Egerton, R.F. (2011). *Electron Energy Loss in the Electron Microscope*, 3e. Heidelberg Germany: Springer.

Flewitt, P.E.J. and Wild, R.K. (2017). *Physical Methods for Materials Characterisation*. Boca Raton, FL, USA: Taylor & Francis Group/CRC Press.

Goodhew, P.J., Humphreys, F.J., and Beanland, R. (2000). *Electron Microscopy and Analysis*, 3e. Roca Baton, FL, USA: Taylor and Francis Group/CRC Press.

Spool, A.M. (2016). *The Practice of ToF-SIMS*. New York, USA: Momentum Press.

Vickerman, J.C. and Briggs, D. (2001). *ToF-SIMS Surface Analysis by Mass Spectrometry*. Manchester, UK: IM Publications, Chichester West Sussex UK/ Surface Spectra.

Vickerman, J.C. and Gilmore, I.S. (2009). *Surface Analysis: The Principal Techniques*. Chichester, UK: Wiley.

# Appendix 1

## Auger Electron Energies

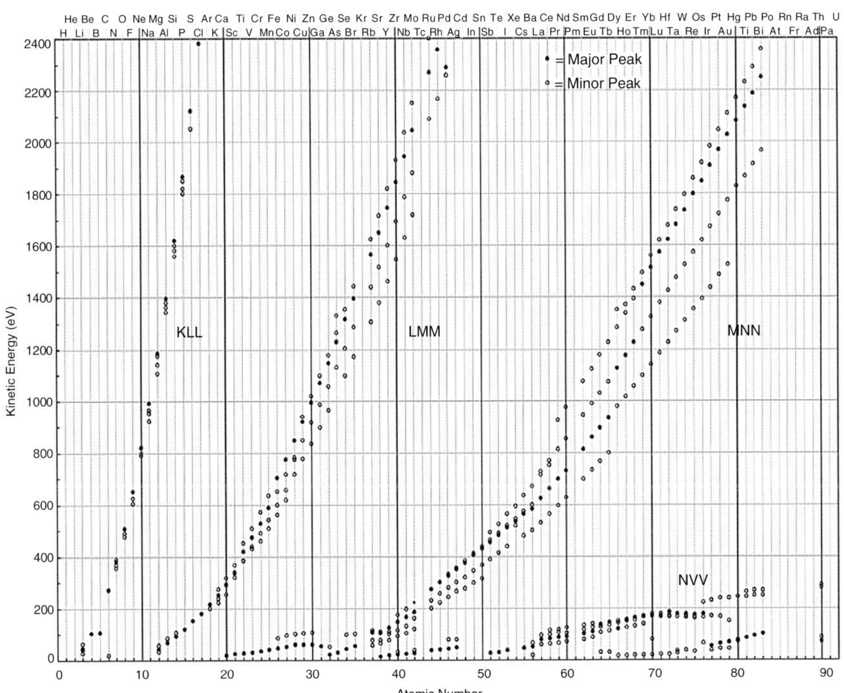


*An Introduction to Surface Analysis by XPS and AES*, Second Edition.
John F. Watts and John Wolstenholme.

# Appendix 2

*An Introduction to Surface Analysis by XPS and AES*, Second Edition.
John F. Watts and John Wolstenholme.

# Table of Binding Energies Accessible with Al Kα Radiation

| Z | | 1 s | 2 s | $2p_{1/2}$ | $2p_{3/2}$ | 3 s | $3p_{1/2}$ | $3p_{3/2}$ | $3d_{3/2}$ | $3d_{5/2}$ | 4 s | $4p_{1/2}$ | $4p_{3/2}$ | $4d_{3/2}$ | $4d_{5/2}$ | $4f_{5/2}$ | $4f_{7/2}$ | 5 s | $5p_{1/2}$ | $5p_{3/2}$ | $5d_{3/2}$ | $5d_{5/2}$ |
|---|---|---|---|---|---|---|---|---|---|---|---|---|---|---|---|---|---|---|---|---|---|---|
| 1 | H | 14 | | | | | | | | | | | | | | | | | | | | |
| 2 | He | 25 | | | | | | | | | | | | | | | | | | | | |
| 3 | Li | 55 | | | | | | | | | | | | | | | | | | | | |
| 4 | Be | 111 | | | | | | | | | | | | | | | | | | | | |
| 5 | B | 188 | | 5 | | | | | | | | | | | | | | | | | | |
| 6 | C | 284 | | 7 | | | | | | | | | | | | | | | | | | |
| 7 | N | 399 | | 9 | | | | | | | | | | | | | | | | | | |
| 8 | O | 532 | 24 | 7 | | | | | | | | | | | | | | | | | | |
| 9 | F | 686 | 31 | 9 | | | | | | | | | | | | | | | | | | |
| 10 | Ne | 867 | 45 | 18 | | | | | | | | | | | | | | | | | | |
| 11 | Na | 1072 | 63 | 31 | | 1 | | | | | | | | | | | | | | | | |
| 12 | Mg | 1305 | 89 | 52 | | 2 | | | | | | | | | | | | | | | | |
| 13 | Al | | 118 | 74 | 73 | 1 | | | | | | | | | | | | | | | | |
| 14 | Si | | 149 | 100 | 99 | 8 | 3 | | | | | | | | | | | | | | | – |
| 15 | P | | 189 | 136 | 135 | 16 | 10 | | | | | | | | | | | | | | | |
| 16 | S | | 229 | 165 | 164 | 16 | 8 | | | | | | | | | | | | | | | |
| 17 | Cl | | 270 | 202 | 200 | 18 | 7 | | | | | | | | | | | | | | | |
| 18 | Ar | | 320 | 247 | 245 | 25 | 12 | | | | | | | | | | | | | | | |
| 19 | K | | 377 | 297 | 294 | 34 | 18 | | | | | | | | | | | | | | | |

| | | | | | | | | | | | | | | |
|---|---|---|---|---|---|---|---|---|---|---|---|---|---|---|
| 20 | Ca | | 438 | 350 | 347 | 44 | 26 | | 5 | | | | | |
| 21 | Sc | | 500 | 407 | 402 | 54 | 32 | | 7 | | | | | |
| 22 | Ti | | 564 | 461 | 455 | 59 | 34 | | 5 | | | | | |
| 23 | V | | 628 | 520 | 513 | 66 | 38 | | 2 | | | | | |
| 24 | Cr | | 695 | 584 | 575 | 74 | 43 | | 2 | | | | | |
| 25 | Mn | | 769 | 652 | 641 | 84 | 49 | | 4 | | | | | |
| 26 | Fe | | 846 | 723 | 710 | 95 | 56 | | 6 | | | | | |
| 27 | Co | | 926 | 794 | 779 | 101 | 60 | | 3 | | | | | |
| 28 | Ni | | 1008 | 872 | 855 | 112 | 68 | | 4 | | | | | |
| 29 | Cu | | 1096 | 951 | 931 | 120 | 74 | | 2 | | | | | |
| 30 | Zn | | 1194 | 1044 | 1021 | 137 | 87 | | 9 | | | | | |
| 31 | Ga | | 1298 | 1143 | 1116 | 158 | 107 | 103 | 18 | | | 1 | | |
| 32 | Ge | | 1413 | 1249 | 1217 | 181 | 129 | 122 | 29 | | | 3 | | |
| 33 | As | | | 1359 | 1323 | 204 | 147 | 141 | 41 | | | 3 | | |
| 34 | Se | | | | | 232 | 168 | 162 | 57 | | | 6 | | |
| 35 | Br | | | | | 257 | 189 | 182 | 70 | 69 | 27 | 5 | | |
| 36 | Kr | | | | | 289 | 223 | 214 | 89 | | 24 | 11 | | |
| 37 | Rb | | | | | 322 | 248 | 239 | 112 | 111 | 30 | 15 | 14 | |
| 38 | Sr | | | | | 358 | 280 | 269 | 135 | 133 | 38 | 20 | | |
| 39 | Y | | | | | 395 | 313 | 301 | 160 | 158 | 46 | 26 | | 3 |
| 40 | Zr | | | | | 431 | 345 | 331 | 183 | 180 | 52 | 29 | | 3 |
| 41 | Nb | | | | | 469 | 379 | 363 | 208 | 205 | 58 | 34 | | 4 |

| Z | | 1s | 2s | $2p_{1/2}$ | $2p_{3/2}$ | 3s | $3p_{1/2}$ | $3p_{3/2}$ | $3d_{3/2}$ | $3d_{5/2}$ | 4s | $4p_{1/2}$ | $4p_{3/2}$ | $4d_{3/2}$ | $4d_{5/2}$ | $4f_{5/2}$ | $4f_{7/2}$ | 5s | $5p_{1/2}$ | $5p_{3/2}$ | $5d_{3/2}$ | $5d_{5/2}$ |
|---|---|---|---|---|---|---|---|---|---|---|---|---|---|---|---|---|---|---|---|---|---|---|
| 42 | Mo | | | | | 505 | 410 | 393 | 230 | 227 | 62 | 35 | | 2 | | | | | | | | |
| 43 | Tc | | | | | 544 | 445 | 425 | 257 | 253 | 68 | 39 | | 2 | | | | | | | | |
| 44 | Ru | | | | | 585 | 483 | 461 | 284 | 279 | 75 | 43 | | 2 | | | | | | | | |
| 45 | Rh | | | | | 627 | 521 | 496 | 312 | 307 | 81 | 48 | | 3 | | | | | | | | |
| 46 | Pd | | | | | 670 | 559 | 531 | 340 | 335 | 86 | 51 | | 1 | | | | | | | | |
| 47 | Ag | | | | | 717 | 602 | 571 | 373 | 367 | 95 | 62 | 56 | 3 | | | | | | | | |
| 48 | Cd | | | | | 770 | 651 | 617 | 411 | 404 | 108 | 67 | | 9 | | | | | 2 | | | |
| 49 | In | | | | | 826 | 702 | 664 | 451 | 443 | 122 | 77 | | 16 | | | | | 1 | | | |
| 50 | Sn | | | | | 884 | 757 | 715 | 494 | 485 | 137 | 89 | | 24 | | | | 1 | 1 | | | |
| 51 | Sb | | | | | 944 | 812 | 766 | 537 | 528 | 152 | 99 | | 32 | | | | 7 | 2 | | | |
| 52 | Te | | | | | 1006 | 870 | 819 | 582 | 572 | 168 | 110 | | 40 | | | | 12 | 2 | | | |
| 53 | I | | | | | 1072 | 931 | 875 | 631 | 620 | 186 | 123 | | 50 | | | | 14 | 3 | | | |
| 54 | Xe | | | | | 1145 | 999 | 937 | 685 | 672 | 208 | 147 | | 63 | | | | 18 | 7 | | | |
| 55 | Cs | | | | | 1217 | 1065 | 998 | 740 | 726 | 231 | 172 | 162 | 79 | 77 | | | 23 | 13 | 12 | | |
| 56 | Ba | | | | | 1293 | 1137 | 1063 | 796 | 781 | 253 | 192 | 180 | 93 | 90 | | | 40 | 17 | 15 | | |
| 57 | La | | | | | 1362 | 1205 | 1124 | 849 | 832 | 271 | 206 | 192 | 99 | | | | 33 | 15 | | | |
| 58 | Ce | | | | | 1435 | 1273 | 1186 | 902 | 884 | 290 | 224 | 208 | 111 | | 1 | | 38 | 20 | | | |
| 59 | Pr | | | | | | 1338 | 1243 | 951 | 931 | 305 | 237 | 218 | 114 | | 2 | | 38 | 23 | | | |
| 60 | Nd | | | | | | 1403 | 1298 | 1000 | 978 | 316 | 244 | 225 | 118 | | 2 | | 38 | 22 | | | |
| 61 | Pm | | | | | | | 1357 | 1052 | 1027 | 331 | 255 | 237 | 121 | | 4 | | 38 | 22 | | | |

| | | | | | | | | | | | | | | | | |
|---|---|---|---|---|---|---|---|---|---|---|---|---|---|---|---|---|
| 62 | Sm | 1421 | 1107 | 1081 | 347 | 267 | 249 | 130 | | 7 | | 39 | 22 | | | |
| 63 | Eu | | 1161 | 1131 | 360 | 284 | 257 | 134 | | 0 | | 32 | 22 | | | |
| 64 | Gd | | 1218 | 1186 | 376 | 289 | 271 | 141 | | 0 | | 36 | 21 | | | |
| 65 | Tb | | 1276 | 1242 | 398 | 311 | 286 | 148 | | 3 | | 40 | 26 | | | |
| 66 | Dy | | 1332 | 1295 | 416 | 332 | 293 | 154 | | 4 | | 63 | 26 | | | |
| 67 | Ho | | 1391 | 1351 | 436 | 343 | 306 | 161 | | 4 | | 51 | 20 | | | |
| 68 | Er | | 1453 | 1409 | 449 | 366 | 320 | 177 | 168 | 4 | | 60 | 29 | | | |
| 69 | Tm | | | | 472 | 386 | 337 | 180 | | 5 | | 53 | 32 | | | |
| 70 | Yb | | | | 487 | 396 | 343 | 197 | 184 | 6 | | 53 | 23 | | | |
| 71 | Lu | | | | 506 | 410 | 359 | 205 | 195 | 7 | | 57 | 28 | | 5 | |
| 72 | Hf | | | | 538 | 437 | 380 | 224 | 214 | 19 | 18 | 65 | 38 | 31 | 7 | |
| 73 | Ta | | | | 566 | 465 | 405 | 242 | 230 | 27 | 25 | 71 | 45 | 37 | 6 | |
| 74 | W | | | | 595 | 492 | 426 | 259 | 246 | 37 | 34 | 77 | 47 | 37 | 6 | |
| 75 | Re | | | | 625 | 518 | 445 | 274 | 260 | 47 | 45 | 83 | 46 | 35 | 4 | |
| 76 | Os | | | | 655 | 547 | 469 | 290 | 273 | 52 | 50 | 84 | 58 | 46 | 0 | |
| 77 | Ir | | | | 690 | 577 | 495 | 312 | 295 | 63 | 60 | 96 | 63 | 51 | 4 | |
| 78 | Pt | | | | 724 | 608 | 519 | 331 | 314 | 74 | 70 | 102 | 66 | 51 | 2 | |
| 79 | Au | | | | 759 | 644 | 546 | 352 | 334 | 87 | 83 | 108 | 72 | 54 | 3 | |
| 80 | Hg | | | | 800 | 677 | 457 | 379 | 360 | 103 | 99 | 120 | 81 | 58 | 7 | |
| 81 | Tl | | | | 846 | 722 | 609 | 407 | 386 | 122 | 118 | 137 | 100 | 76 | 16 | 13 |
| 82 | Pb | | | | 894 | 764 | 645 | 435 | 413 | 143 | 138 | 148 | 105 | 86 | 22 | 20 |
| 83 | Bi | | | | 939 | 806 | 679 | 464 | 440 | 163 | 158 | 160 | 117 | 93 | 27 | 25 |

| **Z** | | **6s** | **6p$_{1/2}$** | **6p$_{3/2}$** | **6d$_{3/2}$** | **6d$_{5/2}$** |
|---|---|---|---|---|---|---|
| 82 | Pb | 3 | 1 | | | |
| 83 | Bi | 8 | 3 | | | |
| 84 | Po | 12 | 5 | | | |
| 85 | At | 18 | 8 | | | |
| 86 | Rn | 26 | 11 | | | |
| 87 | Fr | 34 | 15 | | | |
| 88 | Ra | 44 | 19 | | | |
| 89 | Ac | – | – | – | – | – |
| 90 | Th | 60 | 49 | 43 | 2 | 2 |
| 91 | Pa | – | – | – | – | – |
| 92 | U | 44 | 27 | 16.8 | | |

| Z | | 1s | 2s | $2p_{1/2}$ | $2p_{3/2}$ | 3s | $3p_{1/2}$ | $3p_{3/2}$ | $3d_{3/2}$ | $3d_{5/2}$ | 4s | $4p_{1/2}$ | $4p_{3/2}$ | $4d_{3/2}$ | $4d_{5/2}$ | $4f_{5/2}$ | $4f_{7/2}$ | 5s | $5p_{1/2}$ | $5p_{3/2}$ | $5d_{3/2}$ | $5d_{5/2}$ |
|---|---|---|---|---|---|---|---|---|---|---|---|---|---|---|---|---|---|---|---|---|---|---|
| 84 | Po | | | | | | | | | | 995 | 851 | 705 | 500 | 473 | 184 | | 177 | 132 | 104 | 31 | |
| 85 | At | | | | | | | | | | 1042 | 886 | 740 | 533 | 507 | 210 | | 195 | 148 | 115 | 40 | |
| 86 | Rn | | | | | | | | | | 1097 | 929 | 768 | 567 | 541 | 238 | | 214 | 164 | 127 | 48 | |
| 87 | Fr | | | | | | | | | | 1153 | 980 | 810 | 603 | 577 | 268 | | 234 | 182 | 140 | 58 | |
| 88 | Ra | | | | | | | | | | 1208 | 1058 | 879 | 636 | 603 | 299 | | 254 | 200 | 153 | 68 | |
| 89 | Ac | | | | | | | | | | 1269 | 1080 | 900 | 675 | 639 | 319 | | 272 | 215 | 167 | 80 | |
| 90 | Th | | | | | | | | | | 1330 | 1168 | 968 | 714 | 677 | 344 | 335 | 290 | 229 | 182 | 93 | 88 |
| 91 | Pa | | | | | | | | | | 1387 | 1224 | 1007 | 743 | 708 | 371 | 360 | 310 | 232 | 232 | 94 | 94 |
| 92 | U | | | | | | | | | | 1439 | 1271 | 1043 | 778 | 736 | 388 | 377 | 321 | 257 | 192 | 103 | 94.2 |

# Appendix 3

## Documentary Standards in Surface Analysis

The provision of documentary standards at the international level is the responsibility of Technical Committee 201 of the International Standards Organisation (ISO TC201), full details of which can be found at www.iso.ch. From an analytical point of view documentary standards are essential in any laboratory which runs a quality scheme for the following reasons:

1) To improve reliability of the analytical results obtained.
2) To reduce the level of skill required to perform routine analysis.
3) Data can be transferred between different analytical laboratories, with a high degree of confidence if all laboratories follow a similar procedure.

### The Scope of TC201

Standardisation in the field of surface chemical analysis. Surface chemical analysis includes analytical techniques in which beams of electrons, ions, neutral atoms or molecules, or photons are incident on the specimen material and scattered or emitted electrons, ions, neutral atoms or molecules, or photons are detected. It also includes techniques in which probes are scanned over the surface and surface-related signals are detected.

### The Purpose of TC201

1) To promote the harmonisation of requirements concerning instrument specifications, instrument operations, specimen preparation, data acquisition, data processing, qualitative analysis, quantitative analysis, and reporting of results.

*An Introduction to Surface Analysis by XPS and AES*, Second Edition.
John F. Watts and John Wolstenholme.

2) To establish consistent terminology.
3) To develop recommended procedures and to promote the development of reference materials and reference data to ensure that surface analyses of the required precision and accuracy can be made.

## International Standards Relevant to Electron Spectroscopies

Refer to https://www.iso.org/committee/54618/x/catalogue/p/0/u/1/w/0/d/0 for the current, complete list of Standards developed by ISO/TC 201. At the time of writing, ISO/TC 201 has published 63 Standards and Technical Reports (TR). Of those, the following 38 documents are relevant to the content of this book.

### Terminology

| | |
|---|---|
| ISO 18115-1 | Vocabulary – Part 1: General terms and terms used in spectroscopy |

### General Procedures

| | |
|---|---|
| ISO 16242 | Recording and reporting data in Auger electron spectroscopy |
| ISO 16243 | Recording and reporting data in X-ray photoelectron spectroscopy |
| ISO 18116 | Guidelines for preparation and mounting of specimens for analysis |
| ISO 18117 | Handling of specimens prior to analysis |
| ISO 18516 | Determination of lateral resolution and sharpness in beam based methods with a range from nanometres to micrometres |
| ISO 20579-4 | Guidelines to sample handling, preparation and mounting – Part 4: Reporting information related to the history, preparation, handling and mounting of nano-objects prior to surface analysis |

### Data Management and Treatment

| | |
|---|---|
| ISO 14975 | Information formats |
| ISO 14976 | Data transfer format |

### Depth Profiling

| | |
|---|---|
| ISO 14606 | Optimization using layered systems as reference materials |
| ISO/TR 15969 | Measurement of sputtered depth |
| ISO 16531 | Methods for ion beam alignment and the associated measurement of current or current density for depth profiling in AES and XPS |

| | |
|---|---|
| ISO 17109 | Method for sputter rate determination in X-ray photoelectron spectroscopy, Auger electron spectroscopy and secondary-ion mass spectrometry sputter depth profiling using single and multi-layer thin films |
| ISO/TR 22335 | Measurement of sputtering rate: mesh-replica method using a mechanical stylus profilometer |

## Electron Spectroscopies

| | |
|---|---|
| ISO 10810 | Guidelines for analysis |
| ISO 13424 | Reporting of results of thin-film analysis |
| ISO/TR 14187 | Characterisation of nanostructured materials |
| ISO 14701 | X-ray photoelectron spectroscopy – Measurement of silicon oxide thickness |
| ISO 15470 | X-ray photoelectron spectroscopy – Description of selected instrumental performance parameters |
| ISO 15471 | Auger electron spectroscopy – Description of selected instrumental performance parameters |
| ISO 15472 | X-ray photoelectron spectrometers – Calibration of energy scales |
| ISO 16129 | X-ray photoelectron spectroscopy – Procedures for assessing the day-to-day performance of an X-ray photoelectron spectrometer |
| ISO 17973 | Medium-resolution Auger electron spectrometers – Calibration of energy scales for elemental analysis |
| ISO 17974: | High-resolution Auger electron spectrometers – Calibration of energy scales for elemental and chemical-state analysis |
| ISO 18118 | Guide to the use of experimentally determined relative sensitivity factors for the quantitative analysis of homogeneous materials |
| ISO/TR 18392 | X-ray photoelectron spectroscopy – Procedures for determining backgrounds |
| ISO/TR 18394 | Auger electron spectroscopy – Derivation of chemical information |
| ISO 18554 | Procedures for identifying, estimating and correcting for unintended degradation by X-rays in a material undergoing analysis by X-ray photoelectron spectroscopy |
| ISO 19318 | X-ray photoelectron spectroscopy – Reporting of methods used for charge control and charge correction |
| ISO 19668 | X-ray photoelectron spectroscopy – Estimating and reporting detection limits for elements in homogeneous materials |
| ISO 19830 | Minimum reporting requirements for peak fitting in X-ray photoelectron spectroscopy |
| ISO 20903 | Methods used to determine peak intensities and information required when reporting results |

| | |
|---|---|
| ISO 21270 | X-ray photoelectron and Auger electron spectrometers – Linearity of intensity scale |
| ISO 24236 | Auger electron spectroscopy – Repeatability and constancy of intensity scale |
| ISO 24237 | X-ray photoelectron spectroscopy – Repeatability and constancy of intensity scale |
| ISO 29081 | Auger electron spectroscopy – Reporting of methods used for charge control and charge correction |

# Index



*An Introduction to Surface Analysis by XPS and AES*, Second Edition.
John F. Watts and John Wolstenholme.

## f